Svalbard Life

Paul Wassmann - Rudi Caeyers

AUTHOR
Paul Wassmann

DESIGN AND LAYOUT
Rudi Caeyers

COVER LAYOUT
Rudi Caeyers

PHOTO COVER
Rudi Caeyers

PAPER
Arctic Volume 130 g

PRINTED AND BOUND BY
07-Gruppen, Oslo, Norway, 2013

TYPE
Rotis Semi Sans 11pt

ISBN
978-82-321-0211-2

PUBLISHER
Akademika Publishing
PO box 2461
NO-7005 Trondheim, Norway
Tel.: + 47 73 59 32 10
Email: forlag@akademika.no
www.akademikaforlag.no

PUBLISHING EDITOR
laila.andreassen@akademika.no

FUNDING
This book has been published with funding from and in cooperation with:

FIRST PUBLISHED
May 2013
© Akademika Publishing, Trondheim 2013
© of the texts: their authors, 2013
© of the photographs: their authors, 2013

This publication may not be reproduced, stored in a retrieval system or transmitted in any form or by any means; electronic, electrostatic, magnetic tape, mechanical, photocopying, recording or otherwise, without permission.

We only use environmentally certified printing houses.

CONTENTS

CONTENTS	7
FOREWORD BY ALBERT II OF MONACO	9
INTRODUCTION	11

I. PRELUDE — 17
- Ultima Thule, the Hyperboreans and Apollo — 19
- Darkness and depression? — 20
- Against the current — 23
- Light and exhilaration? — 24
- Arcticism? — 27
- Aim and motivation — 28

II. THE SVALBARD ARCHIPELAGO AT A GLANCE — 33
- Untouched, silent and incomplete... — 35
- Svalbard and its setting — 36
- Svalbard, Spitsbergen, Spitzbergen or Grumant? — 38
- Mountainous landscapes — 40
- Glaciers — 42
- Fjords — 46
- Coastal zone — 52

III. SVALBARD HISTORY: THE EARLY DAYS — 61
- Willem Barentsz discovers Svalbard — 63
- The whaling frenzy — 64
- Intermezzo with Russian hunters — 67
- Early research expeditions — 68
- Dawn of modern scientific investigations — 71
- Significance of the La Recherche expedition — 72
- Early Norwegian attempts to exploit the Arctic Ocean — 75
- Renaissance sounds from the ice — 76

IV. SVALBARD HISTORY: FROM THE EDGE OF CIVILISATION TO A MODERN HUB — 81
- Hunting — 82
- Tourism — 85
- Mining — 86
- Fishing and whaling — 89
- Svalbard part of Norway and international status — 90
- Confrontations during the WWII interlude — 93
- Post-war Svalbard — 94
- Modern times I: the airport — 97
- Modern times II: Tourism — 98
- Modern times III: Research — 101
- Modern times IV: Bustling urban life — 104

V. NATURE — 109
- Plankton — 110
- Benthos — 122
- Fish — 133
- Marine mammals — 139
- Terrestrial mammals — 148
- Birds — 154

VI. COMMUNITIES: NOW AND THEN — 165
- Longyearbyen — 166
- Pyramiden — 170
- Barentsburg — 174
- Smeerenburg — 179
- Virgohamna — 182
- Ny-Ålesund — 186
- Ny-London — 191
- Svea — 192
- Hornsund — 195
- Hopen — 198
- Bjørnøya — 202

VII. PEOPLE: THEN AND NOW — 207
People: then
- Explorers — 208
- Scientists — 215
- Hunters — 224
- Women — 226
- Miners — 234

People: now — 238
- Odd Olsen Ingerø — 239
- Lena Romanenko — 240
- Roger Jacobsen — 240
- Hans Roar Hansen — 240
- Fiona Danks — 241
- Carlos Duarte — 241

Tove Gabrielsen	242
Ole Magnus Rapp	242
Jan Martin Berg	243
Johnna Holding	243

VIII. SVALBARD LIFE IN ART — 247
- Artists and scientists: antagonists or compatriots? — 248
- François-Auguste Biard — 251
- Barthélemy Lauvergne — 252
- Franz Wilhelm Schiertz — 255
- Hans Beat Wieland — 256
- Witold Lovatelli-Colombo — 259
- Louis Tinayre — 260
- Marius Borrel — 263
- Fridtjof Nansen — 264
- Michaloff Wigdehl — 267
- Gert Jynge — 268
- Heinz Köhler — 271
- Andrei Alekseevich Yakovlev — 272
- Kåre Tveter — 275
- Terje Roalkvam — 278
- Vemund Thoe — 281

IX. CLIMATE CHANGE IN THE ARCTIC: WHAT IS AHEAD OF US — 285
- Warmer and wetter — 287
- Amplification of global climate change — 288
- Sensitivity to environmental contaminants and radiation — 291
- Sea-ice decline is a threat to species — 292
- A warmer ocean changes ecosystems — 295
- The inescapable acidification of the ocean — 296
- Landscapes and animals — 299
- A passion for pristine environments: Arctic tourism — 300
- The infrastructure challenges — 303
- Adaptation to change — 304

X. OUTLOOK AND PERSPECTIVE — 309
- Human development and ecosystem change — 310
- Myths about the climate — 313
- Change to new equilibrium? — 314
- Communicate with and to the people? — 317
- The peoples of the Arctic — 318
- The people of the Arctic — 321
- Socialising losses, privatising profit — 322
- Exploitation, preservation and sustainability? — 325
- Resilience assessments and tipping points — 326
- Multidisciplinary awareness: a precondition for resilience management — 329
- The beginning of the end? — 330
- Confidence in the future: a lost skill? — 333
- The serenity of the mind — 334

XI. FAREWELL TO SVALBARD — 339

Photos credits and captions — 340

List over applied publications — 348

Palais de Monaco

December 2012

For over a hundred years, the Principality of Monaco has forged a unique bond with the Arctic regions.

This bond dates back to my great-great-grandfather, Prince Albert I, who led several expeditions to the Far North in the late 19th and early 20th centuries.

His inquiring and wise mind, a pioneer in oceanography and paleontology, was guided by a love for these white expanses, as much as by an extraordinary lust for discovery, progress and science. His expeditions, conducted in extreme conditions and with means that were still at an embryonic stage, led him, among other places, to Spitsbergen where he admired and photographed the majestic silhouettes, which he helped to map.

A hundred years later, I chose to walk in his footsteps by leading an expedition in 2006, which brought me in turn to Svalbard, where I was able to assess the tragic retreat of the glaciers since the photos taken by my ancestor.

The revelation of this indisputable fragility made me even more determined to maintain and strengthen the Principality of Monaco's commitment in favour of the Arctic.

This commitment involves, in particular, support for Polar research, which since the onset has been at the heart of our approach regarding the Far North. It is through science, in a spirit of shared progress and disinterested knowledge of things, that we can, today as in the past, protect the Arctic, its landscape and fauna, but also its people, more effectively.

Because today the entire Arctic is endangered.

This is the case for these unique landscapes that global warming is devouring day after day.

This is the case for these seas that the melting ice has now given over to human appetites, whether for commercial or mining interests.

This is the case too for an exceptional biodiversity which cannot survive the depletion of its biotope.

And above all this is the case for these indigenous peoples whose ancient cultures are now at risk of dying out, confronted with the growing scarcity of their vital resources, a consequence of climate change and the unifying movement of globalization.

Basically, all these phenomena stem from one cause ; the appetite of human beings in relation to the benefits of Nature and their difficulty in accepting any limits to their hold on the world. And they all have the same consequence - the global weakening of our Planet, whose Poles are the first victims.

At a time where the threat of global warming is becoming more and more distinct, we need to protect the Arctic, an essential thermal regulator, with the utmost attention. Because the melting of its sea ice and glaciers will have, we know, a disastrous acceleration effect on the entire climate.

That is why, despite the distance and in spite of the contrasts, the Principality of Monaco more than ever feels a sense of kinship with these distant lands on which our common future to some extent depends.

And this is also why I feel it is more than ever essential to make as many people as possible aware of the reality of the Arctic, as Paul Wassmann's extremely interesting work does, through the example of one of its most emblematic glacial areas, Spitsbergen. It is by raising the awareness of each one of us, regardless of our geographical situation or origin, that we will be able to change the behaviour patterns that are currently threatening Spitsbergen by intensifying global warming.

Therefore, I very much hope that this book, skilfully combining the various approaches and issues, helps to increase awareness of the beauty and fragility of the Far North which must remain an essential sanctuary for nature and life for us all.

Albert de Monaco

INTRODUCTION

Life in Svalbard, frequently called Spitsbergen, has fascinated Europeans for decades. No other place in the Arctic has been so frequently visited, explored, and exploited since the late 16th century than Svalbard. Today, politicians, tourists, students, and scientists travel to Svalbard in ever-greater numbers. However, it appears that the nature, history, and complexity of the archipelago continue to remain obscure to the majority of travellers, whose visits are generally too short to learn more. We take our 'business-as-usual' activities to the 'wilderness' (as mirrored by the above untitled painting of the terminus of a 'wild' glacier by L. Tinayre; see also page 260). Too often, our visits are characterised by a typical arctic attitude: we arrive, grab for what we are looking for and leave. To stimulate our Svalbard voyages with knowledge and comprehension, this book offers a little more substance to the average traveller that visits Svalbard for a few days or to someone that sits in the calmness of home surroundings and wonders about Svalbard and the Arctic.

I have benefited strongly from visions and inspiration provided by colleagues, journalists, and artists that joined researchers during fieldwork in and around Svalbard. Based upon the desire to transfer knowledge in alternative ways to the traditional pathways of science communication, that are obscure to the layperson, the idea arose to exemplify and illustrate life, change, and resilience in Svalbard in manners accessible to a wide spectrum of people. Accordingly, photographs taken by scientists and artists, photographic and graphic evidence from earlier times, and specimens of art have been merged with short texts. This design approach has developed into a book that should appeal to an inherent and wide-ranging awareness of life in the continuously changing Arctic. The book will fulfill its goal if the imaginary (appealing to the imaginative mind) and brief texts (attracting our intellect) result in an immediate understanding of and intuition for life and change in the Arctic.

It has been said that Vincent van Gogh once stated that a paved road is an efficient way to get from A to B, but that unfortunately no flowers grow along it. This book is not such a paved road, but comprises multidisciplinary challenges, deviations, and meanders in the search for 'flowers'. In order to detect such flowers, the book does not follow a straight line, but wanders, and jumps from one central aspect to another in the periphery. It constantly looks for diversity, ideas, and enjoyment. It considers life! The book bears a resemblance to a collage or a mosaic and is thus not meant to be read from the beginning to end, but rather over time, every now and then, and in a 'vaulting' and 'random' manner. Its composition resembles the musical composition *Pictures from an Exhibition*, by the breathtakingly radical Russian composer Modest Mussorgsky, as the author guides the reader through an extensive gallery of various aspects of life in the High North. Life is the theme song of this 'exhibition'. Hopefully, readers will create their own collage, widen it with their own impressions and observations, and develop their own theme songs that in turn create, in a sense, their own 'Svalbard Life' exhibition.

The author (P.W.) and the designer (R.C.) are connected to the Faculty of Biosciences, Fisheries and Economy, at the University of Tromsø. We both thank our faculty for the opportunity and support to make this book a reality. Specifically, I thank Dean Edel Elvevoll for her continuous, sincere, and keen encouragement. We hope that the core issue of any university, i.e. academically scrutinised and well-balanced knowledge, is provided to an interested public in manners that are comprehensible and stimulating. In today's knowledge-based societies, wise decision-making to strengthen resilience and support sustainability is a crucial goal. In order to ease the accessibility of the nature of life in Svalbard in all its diversity and complexity, strict selections of examples have been applied. Only what are considered the most important facts, events, and circumstances and what visitors may easily observe during their visits are presented, and in a short, condensed manner. Near the end of the book, the focus shifts towards the overarching issues of climate change in the European sector of the Arctic Ocean. Finally, a set of short essays dealing with generic aspects of human life on earth is presented, based upon the perspective that the Arctic provides. To facilitate the readability of the book, direct citations of specific literature have been omitted. However, the literature and websites used in the preparation of this book are listed. Without them, the book would never have become a reality. If any facts are incorrect or misinterpreted, they are entirely the responsibility of the author.

Thanks are due to the artists and museums that provided high quality pictures. Particular thanks are owed to Valery Pisano (Musée océanographique de Monaco) and Ann Kristin Balto (Norwegian Polar Institute, Tromsø) for their extensive support. Christian Wexels Riser provided five bird contributions. The names of the many colleagues and contributors that provided photographs are listed at the end. I thank Akademika forlag for approving the publication and Catriona Turner for correcting and improving the English. The book is based upon the research and outreach activities of the Arctic Marine Ecosystem Research Network, ARCTOS (www.arctosresearch.net), a northern Norwegian network that emphasises interdisciplinary approaches to address large-scale and pan-Arctic questions in marine Arctic oceanography.

It is very rarely that a book such as the present one originates like lightening from a blue sky. In this case, the writing of the book matured over decades and during some 30 scientific cruises that the author made to Svalbard and the Barents Sea, during which ideas, history, impressions, and elements of the High North were accumulated and developed. I had never considered writing a book for the general public before Carlos Duarte stimulated and co-authored the book *Arctic Tipping Points* with me in 2011. His direct and fearless attitude greatly influenced my approach towards outreach and communication. The graphic, photographic, and artistic skills of this book's designer, Rudi Caeyers, gave me new insight into how to unify text with imagery in order to obtain a product that both represents a high standard and achieves the goal of addressing the whole reader, not just their intellect. The impact that Eddy Carmack's science has had on my understanding of the Arctic Ocean has reached a point where I cannot imagine how my thinking would be without it. However, two issues have been most significant for the book. The first was my exposure to holistic and multidisciplinary approaches at school and during years spent working with disabled adults. Thanks are due to the teachers at the Rudolf Steiner Schule Ruhrgebiet (Germany) and to the colleagues, villagers, and patients at Camphill Dorfgemeinschaft Lehenhof (Germany) and Trastad Gård (Norway) for instilling in me the perspective that life and existence are multifaceted, holistic, and multidisciplinary phenomena that are disgraced by the 'straightjackets' of any one-eyed worldview. The second issue was the maturation of a Central European who moved to the rim of the Arctic and lived there for four decades. The north Norwegian Rigmor Moelv was essential in this maturation and her encouragement, inspiration, and educated insight became indispensable. Through her, I encountered the multi-ethnic and rich facets of the most international and variable part of Norway: its High North. For this period of learning and understanding, I thank her, my north Norwegian colleagues and friends, and neighbours and co-villagers from Borkenes, Nordreisa and Tromsø. My life would be much more modest without all of you.

The future has already begun. This book is dedicated to the young generation that hopefully is characterised by vital confidence in the future and a serenity of the mind, as personified by Calvin, Linus, Mathis, and Ani.

Tromsø, 01. 03. 2013
Paul Wassmann

I. PRELUDE

ULTIMA THULE, THE HYPERBOREANS AND APOLLO
DARKNESS AND DEPRESSION?
AGAINST THE CURRENT
LIGHT AND EXHILARATION?
ARCTICISM?
AIM AND MOTIVATION

Ultima Thule, the Hyperboreans and Apollo

When the Greek god Apollo, the god of light, music, and truth, left his shrine at Delphi in autumn, he cleared the way for his alter ego, the god of fertility and wine, Dionysus. The Indo-Germanic god, brought to Greece by the Dorian invasion, was replaced by an Asian-based androgynous fertility god that gradually amalgamated northern rigidity with Eastern laxness and law with ecstasy. Where did Apollo spend the wintertime? He proceeded over Parnassus, leaving the muses to themselves, and moved north to the Hyperboreans who lived at the end of the world, Ultima Thule. The Hyperboreans are a legendary people believed to have lived 'beyond the north wind' (from the Greek *hyper* ('beyond') + *boreos* ('north wind')) at the edge of the world, in a land of unbroken sunshine. There, they enjoyed continuous and perfect happiness. Today, we believe that the Hyperboreans must have been the people of the High North, engirdling the eternal ice of the Arctic Ocean. Apollo was wise to go there in winter. Contrary to the beliefs of southerners, the dark period in the High North is an extremely enjoyable one, characterised by extensive festivities and the good company of friends, colleagues, and fellow northerners. The warmth inside, the illuminated beauty of the residences, and the utmost hospitality compensates for the climatic assault on life outside. Apollo's Mediterranean compatriots were ill advised to miss the opportunity of experiencing the cold, dark, and inhospitable nature of High North in winter. Culture counts as much as nature and only through a good balance between these two antipodes can humans live a resilient, well-adjusted life. Indeed, the culture of the hearts of the people of the High North is Mediterranean in nature: generous, extrovert, flexible, and warm. Apollo still knows which peoples he belongs to during winter. Light, music, and truth all have a protector in the High North.

Johanne Marie Hansen-Krone's painting *Skygge i blått rom* (Shadow in blue space) reflects some of the peculiarities of the High North. An archetypical individual in foetal position is seen against the night sky, engirdled by the constellations of winter. Is it an introverted, childlike Apollo that has headed north to hail his hyperborean compatriots? Above his head hovers the Great Bear, *Ursa major*, one of the most pronounced constellations during the long northern winter. The ancient Greeks' named the entire northernmost region *Arktikós*, the country of the great bear (*arktos* = bear). In addition to *Ursa major*, there is another constellation that refers to a bear: the Little Bear (*Ursa minor*). In *Ursa minor* we find the anchor of the Northern Hemisphere, the Pole Star *alpha Ursae minoris*. The arctic creature *par excellence* in the material world, the polar bear, reflects the celestial constellation of the Great and Little Bear on our planet. A link exists between heaven and earth, between celestial and earthly bears, reflecting the first cause of all things. The axis connects the ideal in the firmament with the reality on earth. *Arktikós* circles around the Pole Star axis as a reflection of Apollo's love for the north. Hansen-Krone perceived his archetypical shadow image against a nightly sky. Long live the polar night! ■

Darkness and depression?

North of the Arctic Circle extensive variation in daylight prevails. At the geographical position of Svalbard, the sun is permanently below or just skimming the horizon for four months, providing two interludes of two months duration each in which the amount of daylight increases or decreases by approximately 25 minutes per day. The year in the north is thus not comprised of smoothly changing seasons, but vividly different states, and life has to adapt to these extreme alterations.

Most people from mid- or southern latitudes are convinced that people living in the High North struggle with the darkness, are depressive and drink too much alcohol. Psychological disorders and a high suicide rate are expected. Nothing could be more wrong! Neither psychological nor psychosomatic syndromes and suicides are more prominent in the High North compared to other regions of the World. An investigation among inhabitants of *bella Napoli* (Naples) and the *Paris of the North* (Tromsø) revealed that the Neapolitans were far more depressed than the inhabitants of Tromsø. Clearly, there is far more to experiencing depression than lack of sunlight and low temperatures. Extensive travelling has revealed that these misconceptions are widespread, almost universal. They are thus culturally inherited prejudices that cannot be changed unless one travels to the north with an open mind.

The lack of sunlight, an essential element of life and one not to be missed by northerners, ignites an internal fire, an inner light. The all-supporting nature that maintains southern cultures in the pan-Mediterranean is double-sided in the High North. Nature not only supports, she also attacks. One has to be prepared for her assaults. A serene, all-supportive landscape is created within the High North: an antithesis to the, at times, hostile exterior. The depressing wintery misery is rather encountered in the greyish, foggy and gloomy lowland regions north of the Alps exposed to the Atlantic low-pressure belt than the Arctic. Nowhere can one freeze more than in the Mediterranean during winter. Those that have experienced living for longer periods of time in the often crystal clear, vibrant, and exorbitantly multicoloured Arctic regions can perfectly well compare it with the grey wintery smoothness and indifference of the mid-latitudes and the humid, freezing-cold bluishness of the Mediterranean.

Depression is neither produced by external darkness nor produced by a less supportive nature. Rather, it is produced within us. It reflects our soul and living conditions. If we take care of both, we can create the preconditions for a good life. In the north, all people are well prepared. Welcome to darkness, producer of the inner light! ∎

Against the current

The cultural avant-garde in the 1930s invented a new lifestyle that still dominates our ideas of what constitutes the 'good life'. The zeitgeist changed from a preferable life engirdled by green pastures, moderate temperatures, cultivated land, and clearly recognisable seasons to a life lived along a warm, rainless coast, adjacent to an oligotrophic sea, and with a never-ending succession of sunny days. The body ideal changed from a stout and plump stature to a slim, sporty figure. The appearance altered from established and bourgeoisie-looking to 'eternally' young. The archetype of eternal youth, well known from many cultures, entered the scene and still clearly dominates our choices today. Psychologically, this refers to the emotional life of adults that remains juvenescent at an adolescent level. Regardless of the costs, we wish to stay young all our life. We try to be slim and sporty, have a dark complexion from sun exposure, and be sexually active until our graves. Today, parents use the same clothes as their children, and grandparents adore the members of The Rolling Stones – now wrinkled 'oldies', wrapped in an aura of rocking, youthful eternity.

The change in the zeitgeist resulted in particular conceptions of the significance of light. The majority thinks that longer nights, grey skies, colder temperatures, and life-supporting rain are at best unattractive, but in reality hard to endure. In autumn and winter we thus have bleak moods and long for brighter skies, semi-arid hinterlands, and a blue, low-productive ocean – so off we fly to Spain, the Caribbean, and California. To the dark north? No thanks!

In contradiction to all imaginable, every year people of the High North look forward to the dark period with great expectations, without the slightest symptom of misery. They cannot live without the life-supporting darkness! The northerners of today live against the current of our times. A burning sun, warm nights, and reduced seasonality can hardly be experienced in the north. Zeitgeist enthusiasts and southerners suffer often from extremely narrow climatological niches that makes them unfit to live successfully in a diverse world. Could it be that our universal contemporary focus upon light produces an internal darkness that through various channels find its way to our emotional plane?

Northerners are at odds with the zeitgeist, but they are equipped with more powerful life strategies. They manage the challenges well. The houses are warm, spacious, and in the middle of the dark period they are decorated for Christmas and the New Year. There are festivals, dances, parties, and banquets for friends, colleagues, and neighbours. Children look forward to the snow because playing without snow is considered tedious. They extensively frolic outside in the dark period. Nature supports, but demands an ability to adapt to the seasons and people in the Arctic are skilled in dealing with a well-developed introvert/extrovert psyche. Part of the delight of empathetically embracing life is reflected in the joy that can derive from celebrating. This love for fire, well known by the rapid 'ignition' by northern people when drinking, and the extrovert nature of northerners is frequently misinterpreted as drunkenness and alcohol misuse by outsiders. Apollo, too, needs a break from light, music, the ethereal muses, and crystal-clear truth. Apollo still loves festivals, and law is fused with ecstasy, and rigidity fused with laxness. At times, he undoubtedly asks his alter ego, Dionysus, to join him in Ultima Thule. Where one finds light, one finds shadow, and vice versa. Who studies the shadow of the zeitgeist and the eternal youth syndromes that haunt our times?

■

Light and exhilaration?

The light period stimulates the extensive use of nature and outdoor life. The midnight sun provides energy and delight. One cannot waste time by sleeping. Light activates, and people in the High North sleep less in summer. Working in the garden around midnight is common and one can visit one's neighbour 'in the middle of the night', for coffee. The short, intensive summer in northern Norway is an adventure of unfolding and life. There, one can 'hear' the grass grow in May and June. Plants and animals have to compress their entire reproductive development into a period of just two or three months, before frost, autumn, and snow announce yet another icy interlude of months. They communicate a message that is clearly recognised by humans in the north: 'Do not wait, live now!'

If one is mesmerised by light and colour there is only one good direction – northwards. In the arctic regions the light is far more colourful than in the sharp sunlight, rapid sunrises or sunsets, and high-angled sun of the south. Light and colour change gradually, and particularly during the early and late dark period in Svalbard, the plethoras of colours are beyond description. The red tones of the midnight sun make the green colours of the vegetation fluoresce. Walking through nature in reddish sunlight in summer is an unforgettable experience. When it comes to colours, hippies (children of the zeitgeist), happy to enjoy psychedelic colours, went into the wrong direction. Further, in winter, as much as during the reign of the midnight sun, young and old alike enjoy the archaic mystery, 'filigree' complexity, and animated vivacity of a genuine companion of northerners since time immemorial – the Northern Lights.

Northerners never weary of enjoying the enchantment of the light season and the miraculous colours, numinous dance, and 'whispering' of the Northern Lights. They speak continuously and exhilaratingly to each other about these experiences. Hence, it is not the unexposed traveller that has the strongest impression of light, colour, and nature in the Arctic, but local people that have seen it so many times. Moreover, it binds them and a few migrants to the north with solid and magic ties to Mother Nature. Light entangles them all inside a non-declared confederacy of accomplices, not easily detectable to southerners. ■

Arcticism?

The editors of the essay collection *Arctic Discourses* coined the term 'arcticism' and they did so in analogy of Edward Said's famous, well-known concept of *orientalism*, which describes how a set of restraining images determines the knowledge we gain from a much more heterogeneous terrain. Arcticism shapes our generic image of the Arctic, as reflected in metaphors and expressions such as 'bitterly cold', 'eternal ice', 'virgin land', 'extreme conditions', and 'untouched nature'. The most charismatic representative of the fauna, the fabled polar bear, has risen from the realms of arcticism. Through these customary keyholes we look into the Arctic domain, and the reader, usually living 'on another planet', becomes familiar with the Arctic through these orifices. He or she can connect to a remote world through cultural binoculars and thus remain far away from the reality that is examined. The fascination for the Arctic is unsurpassed exemplified through the iconographic status of the Inuit (still mostly known as Eskimos) as reflected in a suite of romantic novels and widespread terms (such as kayak, anorak, parka, and igloo) and brands (such as clothing and ice cream). However, the use of these words and conceptions does not comprise a deeper understanding of the world of the Inuit. They continue as icons for our perception. They reflect our arcticism and prevent us from experiencing and observing adequately.

The repetition of metaphors and expressions make the unknown seemingly familiar, but leaves little room for the heterogeneous reality. Arcticism identifies a narrative practice for the Arctic that depends on an understanding based upon 'expectations and experience'. Fridtjof Nansen's graphic from the book *Nord i tåkeheimen* (In northern mists) is deeply rooted in our restrained, Arctic images: the lonely wanderer over the eternal virgin snow under the enigmatic miracle of the Northern Lights – the isolated ego surviving in a hostile environment, with the firm ice below and the crystal clear eternity above. Emanuel Kant's metaphor, 'The star-covered sky over me and the moral law in me', comes to mind. In such a manner, Nansen's wanderer meanders in solitude through our existence. Do we get closer to the Arctic through Nansen's graphic and the ideas and emotions that are triggered in ourselves?

In this book, the attention is directed towards both familiar and unfamiliar aspects of life in the Arctic, with the ultimate goal of transferring knowledge in a non-alienating manner. I attempt to omit emblematic catchphrases and clichés regarding the Arctic. Further, I feel attracted to the vision of Charles Baudelaire, who in 1863 characterised the intellectual at the dawn of modernity as the '*perfect flâneur, ... the passionate spectator, ... in the midst of the fugitive and infinite*'. Indeed, I am a flâneur who has lived my daily life in the environments and backdrop depicted here, but was raised in a different cultural context. I have applied an unavoidably personal cultural perspective, but am also a partaker of the fugitive and infinite in the High North. I am a participating observer, and invite the reader to join me in a 'non-arcticism' perspective on the most prominent feature of the European Arctic sector, the Svalbard archipelago. ■

Aim and motivation

This book identifies elements of life in the Svalbard archipelago, its people, and ecosystems. In our era of climate change, the question is raised as to whether Svalbard profits and suffers from the opportunities the changes may bring. This is a timely question, as nowhere else on the planet are the impacts of climate change greater than in the Arctic. A decisive answer is not provided. Rather, the book aims to raise attention regarding the natural joys of living in the High North and exposes people and policymakers alike to the challenges that humanity is facing in the Arctic. It contributes to a continuous process, to which the reader is invited.

In efforts to omit the expectations and pitfalls of arcticism, the book combines information that traditionally would be presented separately. As the author, I wish to contribute to a more balanced comprehension of the heterogenic nature and the historic complexities of the archipelago. I myself have an immigrant background as one living and working in the Arctic. I do not conceive this region as the 'last frontier' (the common perspective for exploiting a 'wild' territory, extract its resources, and thereafter leave), but as a homeland, alongside with locals. The book is not a consecutive narrative, but a collage of impressionistic, mosaic-like fragments that stimulate visions of life in Svalbard and leave the entire vista to the creativity and imagination of the reader and spectator. When we go to the Far North, we are all embedded in Baudelaire's expression of the *fugitive and infinite*, and this book provides elements that add facets to the experience of a magnificent, overwhelming region. When confronted with the amazement and trepidation that Svalbard's nature can impose, every visitor becomes entangled in his own internal world. The chasm between nature outside and the mirror of our soul within can give rise to stark discords that, over time, will provide modulations of our soul and – as I believe – increase our generic, unfathomable understanding of life.

The ultimate aim of this book is the support of multidisciplinary understanding. As will be deliberated later, multidisciplinary and multifacetted approaches are a precondition to reach sustainability and resilience management, the most essential goal for a promising future. This demands openness, interest in diversity, and an appraisal of the ideal of generic, universal education that was left in favour of specialisation. It left the world entrenched in non-communicating '-isms'. Here, such '-isms' attempt to converse with each other and to us. In doing so, the discourse and this book aim to provide elements of what our time needs most – wholeness. ∎

II. THE SVALBARD ARCHIPELAGO AT A GLANCE

SVALBARD AND ITS SETTING
SVALBARD, SPITSBERGEN, SPITZBERGEN OR GRUMANT?
MOUNTAINOUS LANDSCAPES
GLACIERS
FJORDS
COASTAL ZONE

Untouched, silent and complete...

ARCTIC IMPRESSIONS OF THE LAND

In August, 1858, while cruising in my yacht the Ginevra, of 142 tons, on the coast of Norway, I was induced, by the accounts I received of reindeer and other game to be met with in Spitzbergen, to make a trip across from Hammerfest to that country. It being late in the season before we got there, our stay was very short, and our sport was limited to killing a few reindeer, seals, and Brent geese, and to assisting in the harpooning of one or two walruses, in the boats of a sealing brig, which we fell in with amongst the ice. I however saw enough of Spitzbergen to convince me that wonderful sport, and of a most original description, was to be obtained there by anyone who would go at the proper season, with a suitably equipped vessel and proper boats, manned by a crew of men accustomed to the ice and to the pursuit of the walrus and the seal.
(J. Lamont, 1861)

It was at one o'clock in the morning of the 6th of August, 1856, that after having been eleven days at sea, we came to an anchor in the silent haven of English Bay, Spitzbergen. And now, how shall I give you an idea of the wonderful panorama in the midst of which we found ourselves? I think, perhaps, its most striking feature was the stillness, and deadness, and impassibility of this new world: ice, and rock, and water surrounded us; not a sound of any kind interrupted the silence; the sea did not break upon the shore; no bird or any living thing was visible; the midnight sun, by this time muffled in a transparent mist, shed an awful, mysterious lustre on glacier and mountain; no atom of vegetation gave token of the earth's vitality: a universal numbness and dumbness seemed to pervade the solitude.
(Marquess of Dufferin, 1903)

On the third day, after the ship had left Norway and the dark knife-edge of the North Cape had become blurred in the south, we discerned, far distant in the north, a high and bright land. A jagged row of alps with summits and glaciers, a tremendous mountain plateau covered by ice and snow: this is how Spitsbergen appears when arriving from the south. A marvelous white flamboyance rests over this extensive island landscape: snow, ice, and midnight sun over four months of the year.
(A. Schibsted, 1909, translated)

The landscape conveys an impression of absolute permanence. It is not hostile. It is simply there – untouched, silent and complete. It is very lonely, yet the absence of all human traces gives you the feeling you understand this land and can take place in it.
(E. Carpenter, 1973)

Once in his life a man ought to concentrate his mind upon the remembered earth, I believe. He ought to give himself up to a particular landscape in his experience, to look at it from as many angles as he can, to wonder about it, to dwell upon it. He ought to imagine that he touches it with his hands at every season and listens to the sounds that are made upon it.
He ought to imagine the creatures there and all the faintest motions of the wind. He ought to recollect the glare of noon and all the colors of the dawn and dusk.
(From the poem *The Earth*, by N.S. Momaday)

Svalbard and its setting

Svalbard is an archipelago adjacent to the deep Arctic Ocean, situated between Greenland to the west, Franz Josef Land to the east, and northern Norway to the south (page 284). It consists not only of two major islands, Spitsbergen and Nordaustlandet, but also a myriad of smaller ones (page 34). Svalbard is situated between latitude 74° N and 81° N, and from the main settlement, Longyearbyen, there is a distance of only 1300 km to the legendary North Pole, which accompanies us throughout this book. The archipelago has a land surface of about 61,000 km², which is approximately one and a half times larger than Denmark or Switzerland. Glaciers cover 60% of the main islands. Despite the high latitude, the western sections of the archipelago can be reached by ship almost all year round. At no other place on earth can one sail so far north without icebreaking facilities; this includes the island of Bjørnøya (between Spitsbergen and northern Norway) to Svalbard (page 34).

The Svalbard Treaty made the archipelago Norwegian territory. The name Svalbard does not appear in the original document. When the treaty came into force in

1925, Norway renamed the archipelago 'Svalbard' so that it would appear as though the islands had been Norwegian ever since the mysterious Viking voyage in 1194 to the 'cold rim' (*sval bard*). Most probably, the *sval bard* was the marginal ice zone of the Greenland Sea, not Svalbard. The worldwide colloquial term for Svalbard is Spitsbergen, the name of the biggest island.

Svalbard is part of the pan-Arctic region that encircles the Arctic Ocean. The archipelago of Svalbard, with the world's northernmost permanent civil settlements, is situated in the girdle of sea ice that spreads south during winter and recedes north during ice melt in summer. Human settlement in Svalbard is made possible due to branches of the Gulf Stream carrying warm Atlantic water masses derived from the Gulf of Mexico across the North Atlantic into the Arctic Ocean. One branch follows the shelf break to the Barents Sea south of Bjørnøya and spreads along western Spitsbergen northwards, before entering the Arctic Ocean. An eastern branch enters the Barents Sea south of Bjørnøya and spreads through the Barents Sea in a north-easterly direction. Cold water from the Arctic Ocean flows south-west along eastern Svalbard and sends a tongue of cold water along the innermost sections of the west coast of Spitsbergen.

Only a fraction of the archipelago is open for commercial activities such as mining, whereas most of the region is protected in the form of 12 nature reserves. The Governor of Svalbard is extremely strict about ensuring that people that have neither duties nor accepted plans in these regions do not obtain admittance. ∎

Svalbard, Spitsbergen, Spitzbergen or Grumant?

It can be doubted whether the Vikings really found Svalbard in 1194, but it is known for certain that in 1596 the Dutchman Willem Barentsz discovered and mapped several of the islands that today collectively are called Svalbard. Barentsz coined the name Spitsbergen (with the Dutch element 'spits', which translates as 'pointed', and hence Spitsbergen was spelt with a second 's'). At the time, it was believed that Spitsbergen was a part of Greenland, and from the British side the name Greenland was applied for a long time, primarily for political reasons, as they did not want to acknowledge that the Dutch had discovered a separate archipelago. Later, Pomor hunters from northern Russia visited Svalbard and settled there. They called Svalbard 'Grumant', possibly based on the name Greenland. The name Grumant is no longer common, but one of the former Russian coalmines in Isfjorden is called Grumantbyen. A German visitor to the islands in 1675 introduced the German spelling 'Spitzbergen' and subsequently various foreign writers adopted this form. Thus, the four names reflect the history of the archipelago.

Svalbard has Norwegian jurisdiction and flies the Norwegian flag. However, it neither belongs to the European Economic Area nor is it covered by the Schengen Agreement. It has a governor (*sysselmann*) as the representative of the Norwegian state. ■

Svalbard Life | 39

Mountainous landscapes

Svalbard is a mountainous archipelago. In some places, such as in the fjord Hornsund, the mountains appear alpine and spectacular. In other places, the territory is flatter. In still other places, the mountains look like gigantic layers of rock. Seen from above, the mountains of Svalbard appear characteristically flat, and their plateau tops were formed during the last glaciation.

The geological background of Svalbard is most interesting. It is a geologist's Eldorado. Most of bedrock lays exposed. Rocks dating back to most time periods in the earth's history are present in Svalbard. Furthermore, one can find fossils dating from various periods, and spectacular finds of dinosaur footprints and even entire skeletons of dinosaurs have featured in the news in recent years. ■

Glaciers

Most of the ice on earth is not found in the ice-covered seas, but on land in the form of glaciers. In the Arctic, this implies first and foremost the Greenland ice cap, from which enormous amounts of freshwater are discharged into the sea during each summer and early autumn. In concert with river water, melting glaciers are among the most important sources of freshwater and water stratification of the world's oceans. Any changes in climate will immediately be transferred to the cycle of snow precipitation and ice formation and melting that finally are the base of glacier formation. Climate signals are transferred to glaciers, which thus can be used as indicators of climate change.

Glaciers from the interior ice cap move through the frequently narrow fjords and mountain ranges of the Svalbard hinterland, inevitably towards the sea, but often calving within fjords. With 60% of the archipelago covered by ice, most valleys are filled with a network of glaciers, some of them flowing directly into the fjords. Although Svalbard by no means can live up to the might of the glaciers of Greenland, the archipelago is the largest ice-covered region in Europe, with the largest coherent glaciers. The largest unbroken ice cap, Austfonna, is situated on north-eastern Svalbard, where it covers major parts of the island Nordaustlandet. There, it reaches all the way to the sea, along a 190 km long seafront. The ice-cap glacier Austfonna is comparable in size only with Vatnajøkull on Iceland, the second largest European glacier (also an ice-cap type glacier).

Most glaciers in Svalbard have been retreating since the late 19th century. This is evident in the form of moraine ridges located far away from today's glaciers, new islands in fjords, old photographs, and scouring marks along the mountainsides. The ice melts in the lower parts of the glaciers close to the sea, but in the higher parts the scenarios are more complex. Some glaciers increase in volume, some shrink, and some vary in volume. Austfonna, the largest glacier in Svalbard has increased by 50 cm per year during recent decades, while its lower parts have become thinner. Its total volume on land has now almost reached a state of equilibrium between ablation (removal of snow, ice, or water from a glacier or snowfield) and accumulation, despite the fact that the glacier has retreated. However, the amount of icebergs formed by calving has increased and this has resulted in a net ice loss. Clearly, the dynamics of Svalbard's glaciers are complex, and withdrawal in their lower reaches does not necessarily imply that their volume will decrease and that they will eventually disappear. On the contrary, global warming may facilitate an increase of their volume!

The meeting of a glacier with the open water of a fjord is characterised by the dramatic panorama of the calving front, which can be more than 100m high. Such fronts can be monumental and provide impressive scenes. Global warming results in ice melt and some of the melt dramatically takes place along the glacier front when spectacular cascades of crushing ice slump into the fjords. The angle of a glacier serves as an indicator of how quickly it will flow down from the land and into the sea. Ships have to keep a safe distance of three times a glacier front's height, as dramatic calving events can produce dangerous waves. If a glacier front has recently calved, a fascinating turquoise colour will appears among tones of white, grey, and black.

All non-glaciated regions of Svalbard are affected by permafrost. Only the uppermost layers, the active layer, will thaw during the short summer months. Below the active layer, the ground remains frozen all year round. Thawing and freezing of the active layer gives rise to specific surface phenomena, such as pingos (ice munds, covered by soil) and stone rings.

Ice and melting ice are among the most important elements that shape planet Earth. They have affected living conditions since time immemorial. In a warming world and with most people living in regions where ice is rare, there seems to be little understanding of how powerfully ice shapes our lives. Nowhere is the phase change between the crystalline and fluid forms of water so obvious than in the region where glacier and sea ice meet the sea: the fjords. Ice and ice melt impose enormous forces upon land and sea. Simultaneously, we are facing a delicate balance between freezing water and melting ice. Any change in this balance will have significant consequences. ■

Fjords

In common with all mountainous, rocky regions that have been covered by ice during glacial times, Svalbard is characterised by a wide range of fjords, penetrating from the shelves deep into the largest islands, namely Spitsbergen and Nordaustlandet. Fjords are crafted and shaped by glaciers over geological time. They follow ice-flow lines that stretch from the interior of the islands towards the open coast, i.e. they are usual perpendicular to the coastal line. The fjords are the meeting place between the open sea, sea ice, and land. They are protected from the meteorological hardships along the open coast, and the innermost sections are the preferred sites for human settlements. The glorious vistas of the meeting between mountains, glaciers, and the sea make fjords a charismatic goal for the experience of strong contrasts, adventures and tourism.

Fjords can be narrow and embody dramatic, mountainous hinterlands. They can also be wide and open, and encompass a meeting point between an open sea and a high sky. They provide shelter and can be seas of tranquillity, they can be wide-ranging plains of snow, or sometimes hostile locations, covered by rotten ice and drifting ice floes. At other times, they can be whipped into froth and havoc by katabatic winds.

The surface water of fjords is influenced by Arctic water that engirdles Svalbard. However, below this cold water, warmer Atlantic water penetrates into the fjords, carrying with it an Atlantic fauna, namely cod. Thus, the fjords contain a mixture of both Arctic and Atlantic species. In the inner parts of fjords such as Isfjorden and Hornsund, deeper basins can contain cold-water forms even during summer. Arctic species can overwinter there, separated by the fauna of the warmer shelf waters. During periods of warming, the fauna outside the fjords change and Arctic populations become isolated within the fjords. In present-day Spitsbergen – exposed to warming and Atlantic water – Arctic species may become isolated permanently in the innermost basins, as the water outside warms and the range of boreal species expands northwards. Manifold Ice Age relicts, isolated for about 13,000 years, can be found in fjords along the Norwegian coast, dating from times when the ice cap covered the entire Scandinavian Peninsula. The biota in the Arctic Ocean reflects the northwards expansion of boreal species during warming and their exclusion durring cooling. There is continuous competition between the transgression and regression of warmer water species. This makes the Arctic Ocean biota entirely different from those in the seas adjacent to the Antarctic continent, characterised by exceptional environmental stability.

All fjords in the Arctic undergo dramatic functional changes during the course of a year. In late autumn, winter, and spring, i.e. for most of the year, they are part of the open coastal waters. They act as embayments of the open ocean. In summer and early autumn they change into regions where freshwater and low-saline waters are transported in a mighty current towards the open sea, while a compensating subsurface current moves from the sea towards the glacial front. Freshwater from the bottom of the glaciers (resulting from heat generated by friction) is discharged into seawater and rises with considerable speed, jet-like towards the surface. At such times, the water in front of a glacier can 'boil'. Plankton and small fish get out of control by the upwards jet and sudden drop in salinity, and become easy prey for abundant sea birds feeding at the glacier front. Hence, one should watch out for flocks of birds that frequently gather at the fronts of calving glaciers, where most of the freshwater is discharged into the bottom of fjords.

Spectacular sea bird colonies exist along many fjords. On high rocks and mountains, birds seek protection from predators when the route to their feeding grounds is short. In early summer, flocks of birds frantically catch food for their chicks and the sounds of seabirds fill the skies. Fluorescent green patches below the biggest colonies are a telltale sign of poor soils and the beneficiary influence of guano on plant life. Furthermore, such regions are often the preferred grazing grounds of reindeer.

All along the lengths of the fjords one can observe different types of moraines from the various stages of ice retreat. The bigger ones are inundated by seawater and can be found on or off the continental shelves. The recent retreat of the glaciers has exposed clearly visible, fresh side moraines. The van Mijenfjorden is almost closed by a terminal moraine in the form of an island, and the lack of warmer water supplies from the shelf has resulted in the water and fauna within Van Mijenfjorden being quite different from those in other fjords (page 34). ∎

Coastal zone

The coastal zone around Svalbard is a sensitive and important interface between land and sea – a region that contains essential Arctic ecosystems to the east and strongly influenced Atlantic ones to the west. Whereas several coastal zones engirdling the Arctic Ocean face expanding infrastructure investment and growing security concerns, the low number of inhabitants in Svalbard and dedicated environmental protection render increased exposure to hazards less likely. The mighty, snow-covered mountain ranges of west Spitsbergen rise above the horizon as one arrives by ship from the south or from the extensively ice-covered waters of the Fram Strait. First, the highest mountains appear as a ribbon-like Fata Morgana above the horizon, but then gradually during the approach they transform into an awe-inspiring mountain range. Wide vistas allow the eye to search for solid details after the vastness and monotony of the open ocean and ice-covered waters.

The regularly ice-free, relatively warm waters of western Spitsbergen, remotely controlled by the North Atlantic Current (a continuation of the Gulf Stream that originates in the Gulf of Mexico) are frequently covered by mist, fog, and clouds. However, the fjords and the ice-covered waters often experience sunshine and blue skies. This can be observed from far away by the brightly radiant, white horizon, promising that clear skies and the sun are just a few hours away after the expanse

of grey misery at sea. The mixing of cold and warmer waters of the surface and air masses of the atmosphere support the formation of fog, especially during spring and summer.

Along with the relatively warm and fertile waters of the North Atlantic Current boreal species, such as cod, are translocated northwards, spreading southern species far into the north and creating a productive continuum that stretches from the Norwegian coast into the High Arctic. The waters south-west and south of Spitsbergen, and in particular the open shelves and the shelf break, are legendary for their extensive fertility. They support one of the richest fisheries in the world. These waters are, by the grace of the relatively high temperatures, ice-free throughout the winter. Closer to the shore, a band of less fertile Arctic water dominated by arctic species, engirdles Spitsbergen. These coastal waters freeze during winter and much of the ice is transported in a conveyer-belt fashion along with the Arctic water that rounds the southern tip of Spitsbergen or the banks around Spitsbergen. Thus, the coastal zone is exposed to two forms of ice: ice produced in fjords and drift ice from the Barents Sea.

One of the most characteristic, dynamic, and magical regions in the coastal zone around Svalbard is the marginal ice zone where ice cover meets open water. Between the European Arctic region and the North Atlantic Ocean very little landfast ice develops in the coastal zone and an open water marginal ice zone exists all year round, with an ice extent maximum in April and a minimum in October. When moving from open water into the marginal ice zone, visitors are often greeted by bands of melting ice floes, along with a band of light ranging from white to pink in colour on the horizon, setting a mark on the greyish monotony of the open Arctic waters. Sea ice appears most often at the surface, although screw-ice creates ridges at times. However, the subsurface of sea ice is rarely smooth, but is interrupted by cracks, pressure ridges, and pale blue pools of meltwater. It is characterised by a rather complex underwater topography of ridges, caverns, and keels.

The marginal ice zone is also a meeting place in the vertical, between two functionally different regions. Ice and snow cover prevent adequate availability of light and therefore plants do not grow vigorously. However, when snow and ice melt, light penetrates and ice algae grow below the ice. When it melts more and water that has spent the winter trapped under the ice becomes exposed to full sunlight, a remarkable and recurrent bloom of microscopic plants develops, colouring the water slightly green. The marginal ice zone is also an important place for animals. Not only does the bloom of microscopic plants create a signal of production and good feeding conditions for many marine plankton organisms, it is also the resting site for a multitude of marine mammals, such as seals. They may give birth to their pups there and supply them with extremely fat-rich milk for a few days, and they may shed their fur. This region is also the feeding ground for the invincible polar bear, the charismatic creature of the Arctic par excellence, using the marginal ice zone in its full breadth, eventually several hundred or even thousand kilometres away from land. Not only do polar bears hunt seals there, but early in spring this was also an environment exploited by seal hunters, which were engaged in a major Arctic industry until the late 20th century. ∎

III. SVALBARD HISTORY: THE EARLY DAYS

WILLEM BARENTSZ DISCOVERS SVALBARD
THE WHALING FRENZY
INTERMEZZO WITH RUSSIAN HUNTERS
EARLY RESEARCH EXPEDITIONS
DAWN OF MODERN SCIENTIFIC INVESTIGATIONS
SIGNIFICANCE OF THE LA RECHERCHE EXPEDITION
EARLY NORWEGIAN ATTEMPTS TO EXPLOIT THE ARCTIC OCEAN
RENAISSANCE SOUNDS FROM THE ICE

Willem Barentsz discovers Svalbard

SEARCHING FOR THE RICHES OF THE FAR EAST

Svalbard does not enter the realm of history in a sluggish and little-by-little manner, but with drums beating and trumpets sounding. After the discovery of America and the sea route to India and the Far East, the 16th century was ruled by the global expansion of Portugal and Spain. Enormous riches were accumulated, for which other important nations competed, such as the Netherlands, probably the most developed part of Europe. The Netherlands based its wealth on overseas trading. How was it possible for it to participate in the developing markets of India and the Far East? Were there alternative routes to the Far East, uncontrolled by Portugal and Spain through their fortresses along the coasts of Africa and South-America? Could a way be found through either the Northeast Passage or the Northwest Passage, i.e. along the outskirts of the Arctic Ocean?

In 1594 the Netherlands sent their first fleet towards the Northeast Passage under the leadership of the cartographer and navigator Willem Barentsz (1550–1597). Barentsz had travelled widely to the Mediterranean to complete an atlas of this region. He reasoned that an open-water route had to exist north of Siberia in summer because the sun shone continuously and would melt the ice there. Fours small ships entered the Kara Sea. Barentsz reached the west coast of Novaya Zemlya, turned north, but then returned after reaching large amounts of ice. Those aboard had observed polar bears and had succeeded in catching one, but were unable to keep the animal alive when it rampaged the ship. They also came across many walruses, but found it difficult to kill them. This first expedition was considered a success.

Barentsz accompanied the second venture in 1595, that time as the chief pilot and conductor of the expedition. He had a fleet of six ships, loaded with wares that the ever-optimistic Dutch hoped to trade in China! Eventually, the fleet had to return when it was discovered that the Kara Sea was frozen. The second expedition was thus considered a failure.

In 1596, the Dutch Government did not wish to subsidise further expeditions, but instead offered a high reward for anyone that successfully navigated the Northeast Passage. Two ships took a more northern route and discovered Bjørnøya and the west coast of Spitsbergen, where they carried out extensive mapping of the newly discovered land. Barentsz paved the ground for many geographical terms in Svalbard. After returning to Bjørnøya, the two ships took separate routes and Barentsz continued towards the northern region of Novaya Zemlya, made the fatal mistake of turning south on the Kara Sea side and inevitably became stuck in ice floes. Once stranded, the crew was forced to overwinter there. They built a timber lodge, *Het Behouden Huys* (The Saved House), and struggled to survive in the cold; fortunately, the ship had sufficient provisions. The following summer, the crew, their ship still ice-bound, took two rowing boats over the ice to open water and attempt to head south. Willem Barentsz died in misery at sea on 20 June 1597. Almost 240 years later, Christiaan Julius Lodewyck Portmann painted *The Death of Barentsz* (Maritime Museum, Greenwich) in the melodramatic style of that time. After a further seven weeks, the crew reached the Kola Peninsula, where a Russian merchant vessel rescued them. The majority of the team's members reached Amsterdam by November 1597. Thus ended the life of the great explorer Willem Barentsz.

However, Svalbard was suddenly on the map of Europe. The Dutch gave up the idea of navigating through the Northeast Passage, an endeavour first completed by the Finn Adolf Erik Nordenskiöld (1832–1901) on the Swedish ship the *Vega* during the period 1878–1880. In 1853 the former Murmean Sea was renamed the Barents Sea in honour of this great navigator. The Soviet Union made serious attempts to make the Northeast Passage traversable and to develop northern Siberia from the coast inland, among other enterprises by building some of the most powerful atomic icebreakers. Today, following the recent decline in Arctic Ocean sea-ice cover and thickness (an effect of global warming) plans of using the Northeast Passage regularly in the summer to connect with the Far East are once again seriously considered. This new sea route has already started to be used: some test cargo vessels and oil tankers have used the route in the last two years. The future has already begun and early Dutch dreams seem finally to have turned into reality, with the support of anthropogenic forces on the climate. ■

The whaling frenzy

AN EARLY OUTBURST OF ACTIVITY IN THE EUROPEAN ARCTIC

In the 15th century, Western Europe paved the ground for the world power that it was shortly to become. Europe's emerging political and economic growth, reflected in the Dutch expeditions in search of the Northeast Passage, formed the backdrop for the sudden appearance of Svalbard as significant in the European economy. This was due to whaling, particularly the hunt for the slow-swimming bowhead whale, which at the time was abundant around Svalbard. Rivalry among several nations led to efforts to control Arctic whaling grounds and the colossal revenue they represented. Several nations were interested in whaling, including England, Denmark-Norway (in personal union), the Netherlands, and Spain.

Barentsz and his crew had reported extensively on animals they had discovered on their voyages, among them the abundant whale populations of the Svalbard archipelago. Already in 1614 the Dutchman Jan Jacobszoon May van Schellinkhout had come across the island Jan Mayen, subsequently named after its discoverer, and reported an abundance of whales in the region. In these days the lamps of Europe were fuelled by whale oil and there was an immense market for more light. Jan Mayen, on the border between the North Atlantic and the Arctic Oceans, immediately became a base for whalers and periodically up to 1000 men populated the island in the heyday of whaling. In those days, all processing of whales was done at land-based whaling stations.

The whaling adventure in Svalbard started soon after. It is a tale of courageous men, strong competition for resources, the fight for land-based whaling stations, and the large-scale extinction of mammals. The numbers of human casualties caused by capsized rowing boats, scurvy, or other incidents were huge. In those days, everyday life was based upon catching whales with primitive harpoons, flensing the whales on the beaches of shallow bights, and boiling their blubber to extract oil. Such activities continued close to the coast or within the west Spitsbergen fjords until the mid-17th century. The whalers came for a short summer season and the open-air boilers were abandoned in autumn. Over the years, more permanent structures in the form of houses and huts were constructed at several stations. At the end of each season, the barrels of oil were ferried out to ships anchored close to shore and transported homeward.

From the mid-17th century onwards, whales were less numerous in the fjords of Svalbard and whaling shifted focus to more open waters and the ice edge. Ships and techniques were changed accordingly. Whales had to be killed at sea and flensed alongside the ships. Their bubbler was then either boiled on-board the ships or transported in barrels to be boiled upon arrival at the homeport. There were approximately 200–300 whaling ships operating in the European sector of the Arctic Ocean by the end of the 17th century.

Over time, the whaling base at Svalbard lost its significance. The European whaling adventure in the Arctic Ocean came to a preliminary end, and the old whaling stations went out of use. All that remains today are a few ruins of huts and houses, the remains of the boilers on the shoreline, and the graves of many that lost their lives during the breakneck endeavours. The bowhead whale stocks never recovered to its original size. Major sites that reflect the heyday of whaling in Svalbard can be found in Smeerenburg, the 'capital' for whaling at the north-western end of Svalbard. There, up to 200 whalers lived in 16 houses, operating 8 blubber boilers to process an annual catch of 750–1250 whales. On Gravnesset, in outer Magdalenefjorden, there is a major burial ground for whalers. In addition, the Polar Museum in Tromsø and the Svalbard Museum in Longyearbyen house manifold artefacts from the whaling period in Svalbard, including whalers' clothing (found preserved in permafrost), which probably are the only samples of everyday clothing to survive from that time.

The painting by Jacob Storck *Walvisvangst* (Whaling, Rijksmuseum, Amsterdam) depicts Dutch whalers off the Svalbard coast, the hunting of a bowhead whale, and crew members' encounters with polar bears on ice floes. ■

Intermezzo with russian hunters

After the period of Western European whaling – roughly lasting through the whole of the 17th century – hunters from north-western Russia began to operate in Svalbard and their activities continued there until approximately 1850. There has been some debate as to when the Russians first visited Svalbard. It has been claimed that they arrived before 1700, but no records of their dwelling sites, hunting gear, let alone encounters with them, exist from the many reports written by early whalers and whaling expeditions. The Russians were not seasonal migrants with food resources imported from western Europe, like the whalers, but they were well adapted to the Arctic and spent the winters in Svalbard; there were more than 70 stations with wooden huts built from driftwood from Siberia. The majority of the Russians were Pomors, who lived in the White Sea area, a region that has played an important role in the development of Russian national-ethnic self-confidence. The Russian North, which had not been degraded by the Asian stench of the Golden Horde, has contributed to the framing of joint-national culture symbols in architecture, popular poetry, and religious beliefs. In contradiction to most of the solitary Norwegian hunters that came later, the Pomors came often in groups. They where mainly interested in walrus products, but they also traded furs and down. In addition, they hunted reindeer, seals, and birds, and collected eggs, particularly for their own nourishment. The fur of polar foxes and, not least, polar bears, served as an important incentive to the hunters to endure the winter months in Svalbard. Recent excavations of former Pomor huts provided large quantities of handcrafted items, indicating that the Russians spent their spare time processing raw materials by turning them into valuable commodities.

The advent of Russian hunters is seen as a response to the increased demand for products of the North and the corresponding increase in trading activities, attempts by Moscow to limit the trading with foreigners on Russian soil, increased levels of control over trading with Siberia, and decreasing populations of seals and walrus in the White Sea and adjacent regions. The competition from Denmark-Norway to the west made them expand northwards. It is thought that the Russians obtained knowledge about Svalbard from English and Dutch traders visiting the White Sea region and from the visit of Peter the Great to the Netherlands at the end of the 18th century. Turning Russia into a European superpower was the tsar's goal and ever since then Russia has tried to be the dominating Arctic nation through hunting, trading, and seafaring, and after 1918 through its military. The first Russian ship in Svalbard was sighted in 1697. In an early phase, expeditions to Svalbard were organised by the hunters themselves, the monasteries, or town merchants. In the later phase, from approximately 1800 to 1850, merchant unions or trade companies were the dominant force behind Russian hunting. The voyage to and from the White Sea, the survival during winter, and particularly the spring months without supplies of fresh food, and attacks by polar bear turned the hunting expeditions into dangerous endeavours. Arctic hunting by Russians was connected both to the Russian fisheries off the coast of northern Norway and the important Pomor trade between Norway and Russia, whereby fish was exchanged for corn and other goods. The Pomor trade was essential to the withdrawal of northern Norway from the trade monopoly imposed by the cities Bergen and Trondheim. It appears that Russian hunting in Svalbard developed in league with the general upturn in Russian seafaring, of which the Pomor trade was just a part.

The important walrus hunts took place on land and ice. Large herds of walruses were massacred, despite the fact that walruses are huge animals (up to 1000 kilograms or more) with extremely thick skin and thus difficult to kill. They rest in herds on flat shores or ice and move on land only with difficulty. There, the hunters would have approached them silently in small boats and attacked and killed the ones closest to the sea. The dead animals would then create a wall, preventing the rest of the herd from escaping. Neither wooden clubs nor guns were used to kill walrus in the early days. Instead, sharpened lances on wooden poles were used to penetrate the skin and thick layers of blubber. The slaughter sites would have turned into veritable battlefields of blood. Today, heaps of scattered skulls and bones serve to remind us of these times. They make us realise that *Homo sapiens* indeed is capable of changing biodiversity and causing the extinction of species. Moreover, this is not a phenomenon of recent times, but an integrated, deeply rooted characteristic of the human existence.

Orthodox crosses commemorate the graves of perished Russian hunters in the Bellsund region in Svalbard. (lithography by B. Lauvergne (see page 252), La Recherche expedition 1838). ■

Early research expeditions

Following Barentsz's expeditions and the major era of whaling, also the British Royal Navy became interested in Arctic waters. Subsequently, Great Britain became the undisputed leading sea power during the 18th century, also in the High North. Henry Hudson (?-1611), one of the most famous English explorers of the Arctic, also searched for the Orient and returned to Svalbard in 1607, before exploring Novaya Zemlya in the following year. Constantine John Phipps' (1744-1792) expedition visited the Svalbard area in 1773, with the purpose of investigating the geography of the area, which was assumed to be dominated by an open polar sea. Just north of Nordaustlandet, the ship's passage was stopped by solid sea ice. Phipps had a 14-year-old midshipman with him, none other than the great Horatio Nelson (1758-1805), whose greatest wish at the time was to present his father with a polar bear skin (the fulfilment of which almost cost him his life). David Buchan (1780-1838) was appointed to search for a navigable route from Svalbard to the Bering Strait by way of the North Pole (which in surprisingly many contemporary maps was depicted as open water). That voyage marked the Royal Navy's last attempt to sail a ship across the Arctic Ocean.

The 19th century witnessed a significant number of scientific expeditions to Svalbard, mainly with the purpose of studying the geography of the archipelago and other adjoining Arctic regions. However, they were also driven by various nations' desire to display national strength. The Norwegian geologist Balthazar Mathias Keilhau (1797-1858) initiated the first systematic research conducted in Svalbard in 1827. He joined the German industrialist Barto von Löwenigh. Keilhau's narratives presented valuable information about the archipelago, while Löwenigh wrote a travel account. In their accounts, both Keilhau and Löwenigh carefully omitted to mention each other! France was responsible for the most spectacular expeditions, as exemplified by the La Recherche expeditions of 1838 and 1839 (which are the focus of the next section). The famous German geographer August Peterman (1822-1878) visited Svalbard in 1868 and argued the case for a German expedition to the North Pole, using expressions such as *national strength* and *national importance*. Several expeditions expressly wanted to use Svalbard as base for attempts to reach the North Pole. In the mid-19th century, when the new theory of past ice ages competed with Darwin's theory of the origin of species as the hottest issue in contemporary natural sciences, outstanding scientists such as Otto Torell (1828-1900) of Sweden went to Svalbard to study the action of glaciers and ice caps. These expeditions were followed by a series of expeditions during the period 1864-1873, led by the famous Swedish-Finnish geologist Adolf Erik Nordenskiöld. Thus, Svalbard was relatively well known by 1880 and many nations – in various manners – had contributed and made the archipelago the best-known region adjacent to the hitherto still unexplored Arctic Ocean. ∎

Dawn of modern scientific investigations

From its base in Le Havre the French navy ship the *La Recherche* headed for the first time northwards to Norway and Svalbard in 1838, and returned in 1839. The goal was purely scientific: to investigate all aspects of nature and culture in northern Norway and Svalbard. The justification for the expedition came from Louis Philippe d'Orléans, who had fled France during the French Revolution and was living in exile. He conducted incognito a voyage – or should it be called an expedition? – to the North Cape in 1795. After he became King of France (Louis-Phillip I), the remarkable trip inspired him to finance a greater expedition to the European Arctic.

The French physician Paul Gaimard (1790–1858), officer and scientist, was head of the expedition to the European Arctic. Many dedicated and prominent scientists – French and others – participated in this first entirely scientific investigation of the High North. Not all scientists were present during all expedition legs. Many participated only in the investigations of northern Norway and Sweden.

The *La Recherche* sailed northwards from Le Havre in June 1838 and called in at Trondheim. From there, the ship sailed to the present-day principal town of Northern Norway, Hammerfest. From Hammerfest, the *La Recherche* sailed to Svalbard and landed in Bellsund on the south-eastern part of Spitsbergen. Following scientific investigations in the region, the ship sailed back to Hammerfest and made excursions along the coast of Finnmark, before overwintering in France in autumn.

In 1839, the *La Recherche* returned and, after stopping at the Færøy Islands, once again headed for Hammerfest. From there, the ship left for the island Bjørnøya and to Magdalenefjorden in north-west Svalbard, returning to Hammerfest at the end of the summer. Some expedition members left the ship and travelled over Kautokeino, Karesuando and Haparanda to Stockholm, and from there to Paris. The remaining members left by ship at Bergen and Christiania (Oslo). The *La Recherche*'s final voyage was to Finnmark in 1840, which included visits to Reykjavik (Iceland) and Archangelsk (Russia) en route.

The La Recherche expedition rang the bell that marked the start of the age of scientific expeditions to the Arctic, which had a very early start in the European sector of the Arctic Ocean. The manifold investigations were painstakingly published and catalogued. A set of volumes was published over many years. Artists were invited to participate in their preparation, among them Auguste Mayer, Barthélemy Lauvergne, Charles Giraud, and François-Auguste Biard, whose brilliant illustrations appear in all volumes. These graphic documents belong to the first detailed representations of the Arctic and Northern Norway. The depictions of the Sami culture and population are unique. In order to study the largest collection of Sami shaman drums one has to go to Paris, as missionaries burnt all others. The paintings in the volumes represent some of the earliest examples of Sami costumes. Thus, the La Recherche findings and graphics are a goldmine and an extremely valuable starting point for scientific investigations of the High North.

The Norske Nordhav expedition is a second example of the modern age of scientific expeditions to the Arctic. The *R/V Vøringen* expedition took place between 1876 and 1878. It was dominated by natural science research and was directed towards the hitherto little investigated Norwegian, Icelandic, Greenland and Barents Seas. The famous Norwegian scientists Henrik Mohn and George Ossian Sars wrote the following in their application to fund the expedition, which was of basic significance for marine research in Norway: '*this sea ... provided our country with its existence as an inhabited and civilised country. If one visits Asia or America on the same latitude one will find only icy deserts that are populated by nomadic tribes.*' The interpretation of salinity and temperature through the use of mathematical methods to calculate currents in the ocean became a milestone in oceanography. The same is true of the exploration of the deep sea and Arctic marine fauna. Approximately 300 species were described for the first time, and the study of plankton organisms by Sars became part of his systematic treatment of the North Atlantic fauna, which is still *the* basic work on plankton taxonomy in these regions. The final results comprised 28 dissertations that were combined in 7 big volumes, published between 1880 and 1901. During times of very limited funds, an expedition painter had been hired, namely Franz Wilhelm Schiertz, and his expedition paintings are discussed later in this book (pages 254–255).

The lithography by B. Lauvergne (see page 252) depicts the corvette *La Recherche* on her first trip to Svalbard, in the Bellsund region in 1838. ■

Significance of the La Recherche expedition

In an innovative, multidisciplinary, international, and extremely modern approach, the scientists of the La Recherche expedition aimed at exact documentation of the conditions in the High North. They sampled plants and minerals, and compiled reports about the weather and natural conditions. During a time when universal knowledge was assumed the grand ideal, natural science of the region could not, of course, remain in isolation because knowledge of human settlements, living conditions, and anthropological studies was essential. Further, the landscapes, sights, vistas, settlements, and locals had to be presented to Europeans who were hungry to obtain more knowledge about their planet. During the two expeditions to Svalbard, oceanographic experiments were conducted en route. Atmospheric studies (by balloon) and glaciological investigations were carried out, in addition to studies of plant and animal life. Some of the expedition members stayed in Finnmark, for further observations during the winter of 1838–1839. These observations were particularly directed towards the northern lights. All this was published with great care and dedication in 16 volumes of text and 5 volumes of plates, with a general account entitled *Voyages en Scandinavie, en Laponie, au Spitzberg et aux Feröe, pendant les années 1838, 1839, 1840 sur la corvette la Recherche*. Alas, today's science literature is a far cry from the complexity, beauty, dedication, and breadth of those days.

High-ranking Scandinavians joined the French researchers. Among them was the Swedish botanist Lars Levi Læstadius (1800–1861), who obtained permission to leave his position as a priest in Karesuando. Today, he is primarily known as the founder of the religious fundamentalist movement Læstadianism. Læstadius presented 6000 Arctic plant samples to Gaimard, and later received an award from the French Academy of Sciences in recognition of his contribution. Xavier Marmier (1809–1892), a renowned French humanist, conducted cultural-historical investigations. He was interested in people's ability to adapt to harsh conditions, which, as a Parisian in northern Norway, Marmier regarded as extraordinary. He studied the school system and the development of people's intellectual capacity under what he considered inhospitable conditions. However, he was most fascinated by the Sami. Marmier criticised the superficial and unfair image that numerous travellers had imposed upon the Sami. He gained greater insight into their lifestyle after he had shared food and lodged with them. Marmier published his own accounts of the journey in his *Lettres sur le Nord*, in which he paints a lively and seemingly unbiased picture of life in the Far North.

With the La Recherche expedition, modernity was introduced to the no-man's-land of Svalbard. The previously unheard volume of artistic representations gave the educated Europeans a clear picture of Svalbard at the dawn of an era dominated by natural sciences. Readers who are interested in the Arctic and visiting Paris are strongly urged to visit the Geological Museum in Jardin des Plantes. There, one can admire four large Svalbard murals by François-Auguste Biard, one of the main works of this artist, based upon the La Recherche expedition.

The lithography by B. Lauvergne (see page 252) depicts scientists from the La Recherche expedition investigating the littoral zone and the terminus of a glacier of Magdalenefjorden. ∎

Early Norwegian attempts to exploit the Arctic Ocean

Norwegians intensified their hunting and exploration activities when the Russians reduced theirs around 1850. Both groups were interested in more or less the same resources and products, but they were organised in different ways. From the 1850s and onwards, research and expeditions became increasingly important. A series of organised expeditions systematically collected scientific data from the European sector of the Arctic Ocean. Skippers from Tromsø and Hammerfest were an essential element in these endeavours, among them Elling Carlsen (1819-1900). He visited the Svalbard region for the first time in 1843. After several years as a skipper of various vessels, in 1864 he went to the Kara Sea to hunt seals. In 1871, Carlsen found the remains of Willem Barentsz's overwintering camp on Novaya Zemlya, from 1696-1697. He recovered approximately 80 artefacts related to the expedition, among them valuable maps that were sold to a Briton, who later sold them to the Rijksmuseum in Amsterdam. Thus, selected items of the expedition returned to the place where they originally came from 270 years earlier!

In 1872, Carlsen was appointed as ice pilot on the steamer, the *Admiral Tegetthoff* (built in Germany, equipped in Trieste, and sailed by a Croatian crew!), for the Austrian-Hungarian expedition organised by Julius von Payer and Karl Weyprecht. After the vessel had been drifting in the ice for one year, the members of the expedition discovered the archipelago of Frans Josefs Land. Then, after yet another winter spent on the ship, they left it to its destiny and travelled on sledges and rowing boats southwards towards Novaya Zemlya. Carlsen was back in Tromsø by September 1874. Throughout his time as skipper and ice pilot he made essential geographic discoveries throughout the Svalbard and Kara Sea region. His many achievements include the first verified circumnavigation of Spitsbergen in 1863. He once again visited Barentsz's overwintering site in 1876. Three years later, he became the guardian of a lighthouse on Skrova, in the Lofoten archipelago. While living there, he was also hired in 1881 as area expert for the first tourist trip to Svalbard. After he was awarded Swedish-Norwegian and Austrian-Hungarian decorations in 1872, in honour of his contributions, he left Skrova in 1894 for Tromsø where he died in 1900. Indeed, Carlsen had lived an extraordinary life.

The bravery of the Tromsø and Hammerfest skippers was central for the exploration of the European sector of the Arctic Ocean, which became increasingly important. In their pursuit for natural resources such as seals and walruses, or minerals such as coal, they charted landscapes, waters, and sailing routes. Although the effects of these endeavours were of limited economic significance, the scientific results were highly appreciated in academic and political circles. The results shed new light upon global issues such as ocean currents, geological history, Arctic flora and fauna, the northern light, climate, and formation of glaciated terrain.

Also, the First International Polar Year (IPY) in 1882-1883 stimulated further research initiatives in the Svalbard area and other Arctic regions. On three occasions in the past 150 years (i.e. every 50 years), scientists from around the world have worked together to organise concentrated scientific and exploring programmes in the polar regions. In each major thrust, or 'year', scientific knowledge and geographical exploration of Svalbard have been advanced, thereby extending the understanding of many geophysical phenomena that influence the global system. Each IPY is a hallmark of international cooperation in science, and Svalbard has always been central in these efforts.

Towards the end of the 19th century the European Arctic had become an integrated part of Europe, after manifold attempts to gain a foothold. Norwegians became the most active in the European Arctic. The experience gained by scientists and governments in international cooperation set the stage for intensified international science collaboration. This is indispensable because the Arctic Ocean is still the least investigated marine region of the world. Although the effects of global warming are felt everywhere in the High North, our understanding of the pan-Arctic realm is insufficient to comprehend the effects fully. In this regard, improvement will need the concerted action of many nations, far more than at present, and with more dedication and thrust. ∎

Renaissance sounds from the ice

When Elling Carlsen discovered the overwintering place of Willem Barentsz on Novaya Zemlya in 1871 he made an astonishing discovery: among the equipment that Barentsz and his crew had left behind was a wooden flute. Thus, a well-preserved Renaissance flute, probably the only one that has survived the test of time, came into our possession. The rich Renaissance flute literature sends shivering sounds into our times, but what exactly did the music sound like? After 400 years of utter silence, the lone flute has created a living connection to the 16th century. From the flute in the ice, an early music project was born.

In general, a veil of mystery shrouds old music notes and the instruments that were employed. What was the precise tone setting of the flute? How was it played, and how the music phrased? A precise copy was made of the Barentsz flute to test its qualities. It produces a beautiful, clear sound with a large resonance that carries very well. The walls of the flute are only 2.5 mm thick and a musician can feel the wood vibrate when playing the instrument. It was probably produced by a professional flute maker, which was an important profession in times when flute playing was considered so important that some towns in the Netherlands had a separate budget for flautists to entertain the general public. Flute music was also danced to when the sea was calm and this was an important activity aboard ships, as it kept the crew in good physical shape at times when they had an otherwise dull and inactive routine. It is said that dancing was mandatory on long voyages and that the daily routine demanded several hours of dancing!

The Barentsz flute must have had an essential role in the members of the expedition's ability to cope with the physical and psychological challenges of overwintering in a badly isolated hut on Novaya Zemlya – it was taken north to entertain and sustain life on the ice. A copy of the Barentsz flute felt the enigma of the north by making it back to the Arctic again – almost! It became one of the central musical instruments of the early music ensemble *Lux Borea*, in which a traverse flute has always been central. One of the ensemble's projects is *Flute in the Ice*. In addition to the touching story of tormented Dutchmen, the flute tells us a story about the Renaissance traverse flute, which little is known about. All music instruments have to be played and the Barentsz flute can reveal part of its musical backdrop in our time – knowing by doing.

Every now and then, small miracles happen. A flute in the ice provides us with a melody and offers us the aura of bygone days. Let the flute be heard … ∎

IV. FROM THE EDGE OF CIVILISATION TO A MODERN HUB

HUNTING
TOURISM
MINING
FISHING AND WHALING
SVALBARD PART OF NORWAY AND INTERNATIONAL STATUS
CONFRONTATIONS DURING THE WWII INTERLUDE
POST-WAR SVALBARD
MODERN TIMES I: THE AIRPORT
MODERN TIMES II: TOURISM
MODERN TIMES III: RESEARCH
MODERN TIMES IV: BUSTLING URBAN LIFE

Hunting

By 1850 Svalbard had become an integrated part of Northern Europe. Before World War I three activities became the important pillars of the archipelago: hunting, tourism, and mining.

Around the year 1850, Norwegians intensified their hunting activities in Svalbard. What animals were hunted and what where they hunted for? The majority of animals were hunted for their fur, but off course also subsistence hunting was carried out. Both Russian and Norwegian hunters were interested in much the same products. Two terrestrial mammalian species inhabit the archipelago: the Arctic fox and the Svalbard reindeer. Attempts to introduce the Arctic hare and the musk ox failed. There are 15 or more types of marine mammals living in Svalbard's marine waters, including polar bears, whales, dolphins, seals, and walruses. All these animals have been the targets of seasonal and overwintering hunters. Sledge dogs are allowed in Svalbard, but not cats. By the end of the 1800s it had become quite usual for Norwegians to spend most of the year in Svalbard. In winter they hunted foxes and polar bears, and in spring they hunted seals. Also birds and eggs were essential in spring and early summer. In autumn, partridge and reindeer were hunted. Each hunter covered large territories that were connected through a network of cabins and sheds. The total number of hunters was limited, about 50 at the most. Hunting reduced the abundance of some animals, such as polar bears, which were efficiently killed through a crate from which the bear would take some bait, which triggered a shot into its head. The record for hunting polar bears was set by the Henry Rudy, who killed 115 of them in one season.

The economic outcome of the hunting activities, important as they were for the individuals, was never of any significance for those living in Northern Norway, let alone the Norwegian economy as a whole. Hunters had to sell their furs on the mainland or to tourists in order for them and their mainland families to subsist and to buy essential provisions (e.g. food, ammunition, and kerosene) for the forthcoming season (usually 10-11 months), and occasionally some luxury items. Legend has it that the hunter Georg Bjørnnes bought a year's edition of a newspaper for the previous year. Every morning he would go out and fetch 'today's', one-year-old paper. Morning coffee, the newspaper, and peace, these are the principle ingredients of a promising day, even in a hunter's austere cabin. The few that hunted professionally – which was a very different activity compared to what today is considered hunting – enjoyed the nimbus of fame, and continue to do so even today. The life of the hunter appears to appeal to deep, emotional layers of human existence; probably connecting us with the archetypical depths of man's earliest existence – a bygone profession that had deep roots down to strata suffocated by increasingly thickening layers of modern life. In northern Scandinavia, many talk about overwintering, many have read books about the hunters, and a few even apply every year to take over one of the surviving huts in order to overwinter in Svalbard.

In Isfjorden, the area visited by most travellers (because of the main settlements Longyearbyen and Barentsburg), there are remains of Norwegian overwintering huts dating from the very earliest activities until today. Although many of the huts lay in ruins, a few are still standing. Kap Wiik is a hunting station that is still in use, with several smaller huts supporting the hunters' use of the terrain. The place has long traditions. Two of the huts are associated with the legendary hunter Arthur Oxaas (1888–1974). A third one was built by another legendary hunter, Harald Solheim, who gave up a career in zoology at the University of Bergen in order to pursue a hunter's life in Svalbard, and still does so today. Kap Wiik illustrates the development from the most modest to a more comfortable living place, reflecting the recent development of life in Svalbard.

The photographs, taken at the Polar Museum in Tromsø, shows a hunters' cabins and what it was like for hunters in Svalbard. The museum is one of the best polar museums, and its exhibitions of unique objects from Norwegian polar research are fused into the ambience of a historic storehouse in the city's harbour. ■

Tourism

Tourism in Svalbard was established surprisingly early. The first travel account was published as early as 1839, by Barto von Löwenigh. Yachting sportsmen and pleasure hunters followed. From the 1880 and onwards, steamers with wealthy tourists arrived. Thousands of tourists visited Svalbard every year during a short and hectic summer. Some of the rather luxurious vessels accommodated an elite group with a special interest in hunting, and in 1897 one of the Spitsbergen trips was announced as 'Sports Route'. The 'sports route' tourists went ashore or took to the ice whenever life became dreary on-board ship. Heavily equipped, they hunted polar bears and reindeer to such an extent that the latter population levels decreased continuously as long as hunting tourism flourished. As early as 1925, reindeer hunting had to be stopped to prevent extinction. Such hunting had been done purely for pleasure, as no meat was taken to the ships. Nevertheless, nothing is so bad that it is not good for something, and the dead animals must have been a great source of food for scavengers, such as the Arctic fox.

Why did Svalbard become such an attractive tourist destination in the late 19th century? Was it the excitement of visiting *terra nullius* – land that belongs to no man? Was it the possibility to hunt wild animals? Was it the exclusivity of the trip? Alternatively, was it a sudden interest in the Arctic? All these possibilities inevitably played a role, but the most attractive goal in Svalbard and the reason to travel was the North Pole expedition, led by Andrée and Wellmann, to which I will return in chapter VII. In the very early days of aviation, the world was aware of the high-tech approaches used by these two explorers to reach the North Pole by balloon or airship. In those days, the 'Cape Canaveral' was Virgohamna in Svalbard, and thousands of visitors went there to inspect the launch site. One may question whether tourism in Svalbard in the late 19th century and early 20th century would have been so intensive if Andrée and Wellmann had started their North Pole expeditions from Greenland, Canada, or Alaska. I believe the answer must be no.

In addition to English voyages, soon German and Norwegian steamers appeared in the flourishing market for tourism. After trips to the North Cape had become tedious and lost their attraction, the wilderness of Svalbard became one of the most attractive destinations for the rich and privileged. Even a hotel in Norwegian dragon style was erected in 1896 for use in the summer months. It was situated close to Longyearbyen airport, on a headland with a Norwegian name that reflects its location: Hotelnesset. The hotel was quite comfortable and in connection with it the first newspaper of Svalbard, *Spitsbergen Gazette*, which was claimed 'the northernmost newspaper of the world'. From Hotelnesset, tourists visited famous places such as Magdalenefjorden (with its graveyard dating from the early whaling period), Virgohamna (the launch site for Andrée's balloon and Wellmann's airship), a whaling station in Grønfjorden (close to present-day Barentsburg) and mining settlements such as Advent City in Isfjorden or Ny-London in Kongsfjorden. In the years 1896 and 1897 the world paid immense attention to Svalbard, due to increased tourism, improved infrastructure, and the fascination of Andrée's attempt to fly with a balloon to the North Pole. The North Pole continues to be a charismatic spot. Some years ago, Russia planted the Russian flag at the bottom of the sea below the North Pole and recently a Russian Orthodox bishop blessed it, seemingly to demonstrate and underline Russia's ownership.

After the outbreak of World War I tourism ceased, and even long after World War II it was never as popular as it had been earlier. Thus, ended 35 years characterised by bustling summer activity, luxury ships, exquisite menus, prosperous bourgeoisie, the upper classes, and land visits that were not celebrated for their environmental caution. After the whaling frenzy in the 17th century, no-man's-land Svalbard had once again become an integrated part of Europe, through tourism. However, Svalbard paid a toll for its touristic visitors. Animal populations decreased, the wooden structures of archaeological sites were used for bonfires and graveyards were mistreated. It is told that many a writing table in Europe was decorated by a 'To be, or not to be' memento in the form of a human skull from the Gravnesset graveyard in Magdalenefjorden. Arctic tourism to remote places has not ceased. A few tourists pay high prices to travel by Russian atomic icebreakers to the North Pole, one of the most desolate locations on earth.

In his book *Greetings from Spitsbergen*, John T. Reilly provides detailed insight into this legendary period that is also known as a time of prolific postcard writing. It is said that approximately 20,000 postcards were written during the course of one cruise, i.e. an average of 50 cards per passenger, with record being more than 100 postcards for one passenger in a single day. The postcards probably caused blisters or even bleeding wounds in the hands of the poor postman that had to place a clearly legible stamp on each of the cards and letters. Ships sometimes ran out of supplies of postage stamps. This was the heyday of photographers, illustrators, and painters of images reproduced on postcards.

Today, the practice of sending postcards is on the decline. Instead, assemblies of mobile phones with a plethora of sound sequences, and tourists communicating loudly, applying an embarrassing range of vernacular languages, have taken over the silent composition of postcards. This is possible only in places where coverage exists, i.e. within and outside Longyearbyen and Barentsburg. Elsewhere, the archipelago is shrouded in 'mobile-phonic' silence, which poses a challenge for many visitors. By contrast, a few others utter sighs of relief.

The graphic by the maritime painter Kenneth D. Shoesmith, who is well-known for his images of ships and posters advertising trips to exotic places, depicts the vessel Atlantis, anchoring in bright sunshine in a wintry Svalbard fjord. ∎

Mining

The most important Svalbard industry over the years has been coalmining. Coal deposits were detected in Svalbard in the 1890s, and Norwegians were responsible for the first mining activities along Isfjorden in 1899. By 1904, British interests had been established in Adventfjorden and the first year-round operations were started. In 1908, coal production started in Longyearbyen, initiated by the legendary American John Munro Longyear (1850-1922), who travelled to Svalbard in 1901, as a tourist with his wife and five children. The educated and experienced man's family cruise led to the establishment of the base named Longyearbyen, after its founder. Longyear was a total abstainer, who used more time in the cruising vessel's library than at the bar. There, he learned about Svalbard and despite being on holiday, he paid attention to the potentials for new businesses. He saw the possibilities for coalmining and this ignited his interest. In 1903 he visited Adventfjorden, analysed the coal and took over an annexation (in those days of no-man's-land, one could annex a region for one's own interest) and started regular mining in 1904. Permanent housing, water supplies, and a cable car were built. From 1906 and onwards people started living in what then was called Longyear City. After 1926, the settlement became known as Longyearbyen.

Store Norske Spitsbergen Kullkompani (SNSK, the Great Norwegian Spitsbergen Coal Company), a subsidiary of the Norwegian Ministry of Trade and Industry, is the most important coal company in Svalbard. The company established itself in 1916, as did other Norwegian interests during World War I, in part by buying American interests. Dutch, Russian, British, Swedish, and Soviet interests were dominant, but after World War I the demand for coal declined and mines were abandoned or sold. With the destruction of mining settlements during World War II, a new start was made in 1946, but only the USSR (Barentsburg, Grumantbyen, and Pyramiden) and Norway (Longyearbyen, Ny-Ålesund, and later Svea) pursued their interests. For more than 100 years, coalmining had generated many work opportunities for the inhabitants of Norway, Russia, and Ukraine, and many families today can tell stories about relatives that went to Spitsbergen to earn money, often under difficult conditions. Many recount that life as a coalminer was hard, but few complain.

Until recently, coalmining was Svalbard's dominant commercial activity. SNSK operates the mine Svea Nord in Svea and Mine 7 in Longyearbyen. The former produced 3.4 million tonnes in 2008. In the case of Mine 7, 35% of the output is used by the Longyearbyen power station. Since 2007, there has not been any significant mining by the Russian state-owned company Arktikugol in Barentsburg. Test drilling for petroleum on land was done, but it did not give satisfactory results to justify a permanent operation. The Norwegian authorities no longer allow offshore petroleum activities, for environmental reasons, and the test-drilled land has been protected as natural reserves or national parks.

Working conditions in the mines and in the settlements were challenging. Within the mines, the coal was generally less than 1 metre thick and the temperatures were a constant -4°C. Until Word War II miners travelled to work along steep tracks on the snow- and ice-covered slopes, in darkness in winter, and exposed to icy winds, if not blizzards. At least until the 1970s, living accommodation in Longyearbyen was spartan. The Soviet propaganda proudly stated that community facilities in Pyramiden and Barentsburg were better than those of the Norwegian settlements. There was even a swimming pool at Pyramiden, which was by far the most northerly pool in the world! The communities were male-dominated, although miners were able to bring their wives with them if they could find work (such as in the laundry or canteen, or as wardens of the hostel blocks). Generally, in the summer the miners returned to mainland Norway or to Russia, and in many cases to Ukraine, for a few weeks, because the main production activity took place in winter. Summertime efforts were concentrated on loading bulk carriers from the huge stockpiles of coal that had accumulated, ready for when the sea ice cleared.

The Norwegian artist Gert Jynge (see page 268) drew this sketch inside one of the Longyearbyen coalmines in 1930. ∎

Fishing and whaling

The waters around Svalbard, particularly to the south and east, are renowned for their high productivity. They support one of the richest and best-managed fisheries in the world and represent a major segment of the Norwegian and Russian economy. Why, then, are fisheries not mentioned as an important economic activity and employment in Svalbard? The reason is that fish is not landed on the archipelago, but transported directly to Russia, Norway, or other markets abroad. Hence, a visitor will rarely see a fishing vessel in the immediate vicinity of Svalbard, let alone in Barentsburg or Longyearbyen.

There have been interludes when fishing in Svalbard fjords had some significance. A warming period in the 1930s supported a cod fishery in Kongsfjorden, which, following the end of mining due to low coal prices, resulted in an attempt to turn Ny-Ålesund into a fishery settlement. However, during the 1940s colder water returned and the cod disappeared. Svend Føyn, a Norwegian whaling and shipping magnate, pioneered revolutionary methods for hunting and processing whales. In 1873 he introduced the modern harpoon gun and brought whaling into the modern age, resulting in the disastrous global depletion of several whale species. Catching fin whales and blue whales off Svalbard became important in the late 19th century and a whaling station was active at Green Harbour, close to Barentsburg. However, this was only a short intermezzo and overexploitation soon resulted in declining populations. Today, Norway annually catches a sustainable quota of minke whales during their migration along the Norwegian coast to Svalbard and the Barents Sea. Sealing has not been of particular importance in recent decades, but earlier it was an important business for several Arctic nations. Given the embargo on all seal products imposed by the European Union, seal hunting is unlikely to pick up again in the foreseeable future, despite the high abundance of most seal species in Arctic waters.

With the current and anticipated climatic and ecosystem changes, new species of economic importance may be exploited in Svalbard waters in the future. However, there is reason to believe that these resources will not have an impact in Svalbard, but on the mainland economy.

This painting by F. Tinayre (see page 260) is from 1907 and depicts the flensing of a fin whale at the whaling station at Green Harbour, an important destination for tourists around the turn of the 19th century. ■

Svalbard part of Norway and international status

The Svalbard archipelago has been Norwegian territory since 1920. The 'Treaty between Norway, the United States of America, Denmark, France, Italy, Japan, the Netherlands, Great Britain and Ireland and the British Overseas Dominions and Sweden concerning Spitsbergen' (commonly called the Svalbard Treaty or the Spitsbergen Treaty) was signed in Paris on 9 February 1920. It recognises the full and absolute sovereignty of Norway over the Arctic archipelago of Svalbard, at that time called Spitsbergen. The exercise of sovereignty is, however, subject to certain stipulations, and Norwegian law does not fully apply. The treaty regulates the degree of militarisation of the archipelago. The signatories were given equal rights to engage in commercial activities (mainly coalmining) on the islands, and as of 2012, Norway and Russia have been exercising their rights.

The treaty was submitted for registration in the *League of Nations Treaty Series* on 21 October 1920. In addition to the 14 original High Contracting Parties, several additional nations signed within the following five years, before the treaty came into force, including the Soviet Union in 1924 and Germany and China in 1925. There are now over 40 signatories. On 14 August 1925, the treaty came into force. Norway then took over sovereign governorship and immediately enacted a series of environmental protection measures. The following are the essential points of the treaty:

- *Svalbard is part of Norway*: Svalbard is completely controlled and part of the Kingdom of Norway. However, Norway's power over Svalbard is restricted to the limitations listed below.

- *Taxation*: This allows taxes to be collected, but only enough to support Svalbard and the Svalbard government. This results in lower taxes than mainland Norway and the exclusion of any taxes in Svalbard supporting Norway directly. In addition, Svalbard's revenues and expenses are separately budgeted from mainland Norway.

- *Environmental conservation*: Norway must respect and preserve the Svalbard environment

- *Non-discrimination*: All citizens and all companies of every nation under the treaty are allowed to become residents and to have access to Svalbard, including the right to fish, hunt, or undertake any kind of maritime, industrial, mining, or trading activity. The residents of Svalbard must adhere to Norwegian law, and Norwegian authorities cannot discriminate or favour any residents of a certain nationality.

- *Military restrictions*: Article 9 prohibits naval bases and fortifications, and the use of Svalbard for warlike purposes. However, the archipelago is not entirely demilitarised. ■

Le présent Traité entrera en vigueur, en ce qui concerne les stipulations de l'article 8, dès qu'il aura été ratifié par chacune des Puissances signataires, et, à tous autres égards, en même temps que le régime minier prévu audit article.

Les tierces Puissances seront invitées par le Gouvernement de la République française à adhérer au présent Traité dûment ratifié. Cette adhésion sera effectuée par voie de notification adressée au Gouvernement français, à qui il appartiendra d'en aviser les autres Parties Contractantes.

En foi de quoi, les Plénipotentiaires susnommés ont signé le présent Traité.

Fait à Paris, le neuf février 1920, en deux exemplaires, dont un sera remis au Gouvernement de Sa Majesté le Roi de Norvège et un restera déposé dans les archives du Gouvernement de la République française et dont les expéditions authentiques seront remises aux autres Puissances signataires.

The present Treaty will come into force, in so far as the stipulations of Article 8 are concerned, from the date of its ratification by all the signatory Powers; and in all other respects on the same date as the mining regulations provided for in that Article.

Third Powers will be invited by the Government of the French Republic to adhere to the present Treaty duly ratified. This adhesion shall be effected by a communication addressed to the French Government, which will undertake to notify the other Contracting Parties.

In witness whereof the abovenamed Plenipotentiaries have signed the present Treaty.

Done at Paris, the ninth day of February, 1920, in duplicate, one copy to be transmitted to the Government of His Majesty the King of Norway, and one deposited in the archives of the French Republic; authenticated copies will be transmitted to the other Signatory Powers.

Confrontations during the WWII interlude

Svalbard remained unaffected by the German occupation of Norway in April 1940 and life continued as before, for both the Norwegian and Russian settlements. The Norwegian government-in-exile rejected the suggestion of a Soviet-British occupation of Svalbard. There were also suggestions by the USSR to make Bjørnøya a military stronghold for the USSR navy. The USSR had a non-aggression pact with Germany until 22 June 1941, but following the German attack on the USSR the archipelago became of strategic importance in the supply chain between the Allies and the USSR. It became also a source of badly needed coal. On 29 August 1941, the entire population of Ny-Ålesund was evacuated to Longyearbyen, and on 3 September, 765 people were evacuated from Longyearbyen to Scotland. Later, the remaining 150 men were evacuated. Also the Russian population was evacuated to the USSR. British and Canadian forces destroyed installations, mainly Soviet coalmines, and prevented the Germans from occupying these settlements. With Longyearbyen depopulated, a small German garrison and airstrip was established in Adventdalen (Advent Valley), mostly to provide meteorological data. After the British troops regained control, the German forces left Longyearbyen without combat.

In September 1943, Germany dispatched battleships and destroyers to bombard Longyearbyen, Barentsburg, and Grumant. Only four buildings in Longyearbyen survived: the hospital, the power station, an office building, and a residential building. Sverdrupbyen, situated in the inner part of Adventdalen and protected by a small hill, also survived. Weather forecasts became essential for the German armed forces to fight the supply convoys of the Allied forces to Murmansk, an essential support for the fighting capacity of the USSR. Local confrontations and fights broke out between German and Norwegian troops in various places in the archipelago during the war, often related to secret German weather stations,

but for the most part Svalbard was not involved in the last two years of World War II. Longyearbyen remained unsettled until the end of the war, with the first ship from the mainland leaving on 27 June 1945.

Norwegian soldiers, sent by the Norwegian government-in-exile in London on a secret patrol to detect German activity in Svalbard, inspect the ruins of Longyearbyen and transport supplies after the settlement was completely destroyed during World War II. ■

Post-war period

Plans were laid during World War II to ensure quick reconstruction and commencement of mining. Longyearbyen developed rapidly, but until the construction of the airport it was accessible only by ship. By 1948, coal production reached the pre-war level. The original centre of Longyearbyen, close to the present-day centre, was only partly reconstructed. A part of Longyearbyen was established in 1946 as Nybyen ('New city') and consisted of five barracks, each housing 72 people. The first issue of the local newspaper *Svalbardposten* was published in November 1948. In 1949, Longyearbyen received telephone service via a radio connection between Svalbard Radio (located just outside Isfjorden) and the mainland. In 1949, a farm was built in Longyearbyen to keep dairy cattle (for milk), pigs, and hens. A local radio station started broadcasting in Longyearbyen in 1950. The burial ground remained in use until 1950. However, it was discovered that the bodies were failing to decompose due to permafrost, and since then all bodies have been sent to the mainland for burial. A community centre named *Huset* (The House) opened in 1951. During that period, the centre of Longyearbyen shifted towards the upper section of Adventdalen. In 1957, a principal was hired at the primary school and a new church was opened in 1958. A bank opened in 1959.

In the 1960s, the farm was closed and its milk production was replaced by industrially produced milk (powder). From 1961, a private middle school was constructed. New mines were successively opened, and later closed when the coal seams had been exhausted or were struck by accidents. The first commercially produced snowmobile – today's icon of personal independence – was taken into use in 1961. Television broadcasting equipment was installed in 1969, with the Norwegian Broadcasting Corporation's scheduled programmes being sent with a two-week delay.

In 1971, a new school building combining a primary school and lower secondary school was opened, along with a new sports hall and a swimming pool. The Svalbard Council was established in 1971. It consisted of 17 non-party members who were elected or appointed by three different groups: Store Norske Svalbard Kullkompani (SNKS) employees, government employees, and others. In 1973, the Ministry of Trade and Industry bought one-third of SNSK. It continued to buy additional shares until it had acquired almost 100% in 1976. In 1978, the community received satellite communication with the mainland. In the same year, an upper secondary programme was introduced at the public school.

From 1984, television programmes were broadcast live via satellite.

This impressive history of development since World War II has gradually connected Svalbard and Longyearbyen more closely with the mainland. For the people living there, it has been a long march. Originally, workers and administrative staff had lived apart, had eaten different types of food, and had followed different rules. Specific SNKS money existed and there was no regular store; rather, all was controlled, regulated, and prescribed by the SNKS, and life resembled that of the socialistic settlements of Pyramiden and Barentsburg. Today, some remnants of that time still survive. One can buy wine ad libitum (formerly a privilege of administrative staff) but one still needs permission to buy beer (which formerly was consumed by the workers)!

A photograph in early autumn from 1950 shows Nybyen (left) and Sverdrupbyen (right). ■

Modern times I: the airport

From the beginning of the time when expeditions that aimed to reach the North Pole, following the failed attempts by Nansen on the *Fram*, Andrée in his balloon, and Wellmann by airship, planes came into play. Early on, seaplanes operated successfully out of Svalbard. The German Air Force often used planes during their operations in Svalbard. The Norwegian Air Force started serving Longyearbyen with postal flights in the 1950s, by making airdrops. In 1959, when a man fell seriously ill, a landing strip was prepared in Adventdalen. In the same year, the Norwegian airline Braathens SAFE started serving the tundra airport in Adventdalen with irregular winter flights. The backbone of communication with Svalbard and Longyearbyen was the coal transport ships and passenger vessels such as Hurtigruten, which operated voyages along the Norwegian coast in summer. The last vessel sailing Longyearbyen departed from Norway in early November (with the last supplies for the winter and Christmas post, presents, and trees) until traffic commenced again when the fjord became ice-free in May or June. Otherwise, Svalbard was more or less isolated. This situation changed dramatically with the opening of the Longyearbyen international airport in 1975. The airport was built despite protests by the USSR due to fears that it could be used for military operations.

The entire community of Longyearbyen was gathered on 2 September 1975 for a gala dinner in the airport hangar and King Olav V of Norway and the members of the Norwegian Government planned to arrive by plane from Tromsø. It was one of these typically grey, miserable and foggy days and despite two attempts to land the plane was forced to return for refuelling to Tromsø, a 180 minutes round trip. Consequently, a great day for Svalbard and Longyearbyen lacked the finishing touches and the presence of the Norwegian sovereign and government representatives. The airport initially provided four weekly services to mainland Norway and semi-weekly services to Russia.

Today, the airport serves also for helicopter rescue operations, internal flights between Svea and Ny-Ålesund, and special planes serving Russian camps on drift ice in the central Arctic Ocean. Flights depart almost daily all year round and the airport was expanded and upgraded a few years ago. All fresh food destined for Longyearbyen is now transported by plane. The number of people that travel to and from Longyearbyen is amazing. Svalbard has become a major holiday destination and is a popular place for weekend trips. Companies, institutions, and politicians hold meetings in Longyearbyen and it seems that the entire Norwegian leadership has regular Svalbard visits. However, this level of interest in the Arctic and frenzy comes at a price: a lot of pressure upon a vulnerable environment (e.g. in terms of water, sewage, litter, and power plant smoke) and high CO_2 emissions (more than 350 kg per person per flight from Tromsø). Thus, modern man celebrates and experiences the Arctic by neglecting the environmental impact. Yet, there is no such thing as a free lunch, and experiencing the wilderness comes at an environmental price. ■

Modern times II: tourism

Since the opening of the main airport in Svalbard, tourism has significantly increased in the archipelago. Also the number and size of tourist ships has increased greatly. Space at the Longyearbyen harbour sites can sometimes be in short supply. The annual number of visitors to Longyearbyen outnumbers the inhabitants by more than 20 times. One has to realise that Arctic environments are fragile, and that increasingly they should be taken care of by the authorities. Tourists are welcome to visit Svalbard as long as their impact is small. Places such as Longyearbyen clearly exemplify how much tourism shapes modern Svalbard. With almost daily flights from Oslo or Tromsø to Longyearbyen, no other Arctic outpost is so easily accessible within a few hours.

Apart from the coalmining industry, the other main activities in Svalbard are tourism and research, and the latter two depend on economic surplus created elsewhere. Why is there all this frenzy for an outpost that is not sustainable? The fascination for Svalbard seems to derive from its wilderness, which is considered to represent one of the last vestiges of a past, dominated by unspoiled nature. The term wilderness brings to mind regions untouched by the modern world. Today, wilderness areas can be promoted as products or sites of consumption, after which they cease to be true wilderness.

Isfjorden in Svalbard no longer lies beyond the last wilderness frontier, and tourism (the world's largest and fastest growing industry) has contributed to this situation. The frontier towards the Arctic has already shifted northwards and – if the current rate of ice melt continues – will soon become obsolete. The wilderness of the

High North has long been considered a fearful and dangerous place, but is now regarded by tourists as one of the most desirable places to visit. Wilderness can only be preserved by not visiting it. Can we convince humans to leave parts of the world untouched, unvisited, and unseen in order to preserve part of nature with its original biodiversity? What about our remote impact on wilderness through atmospheric emissions of contaminants, soot, and gasses? Is it the fate of our planet that even the most remote spots are ultimately transformed from nature to culture? These are rhetorical questions raised in response to modern tourism. The moloch of civilisation demands a costly sacrifice.

For the average tourist, who belongs to by far the greatest fraction of people visiting Svalbard, there is plenty of 'wildernesses' to explore and visit. Ships shuttle visitors back and forth to the interesting sites. Snowmobiles carry visitors to remote places, and there are possibilities of dog sledding. Unless visitors are very well acquainted with Arctic conditions, they should not leave the settlements without an experienced guide. In particular, polar bears represent a danger to be taken seriously. Carrying rifles in public, reinforced fences around kindergartens, and fairly regular reports of polar bear sightings all add to the wilderness character of Svalbard.

For Longyearbyen, Barentsburg and Ny-Ålesund, tourism is an important source of income and one that these communities depend on. This is particularly the case for Longyearbyen, where hotels, restaurants, and shops depend heavily upon cash flow through tourism. For the few that have experienced real Arctic wilderness, the Longyearbyen region has been so strongly impacted by industry for more than 100 years that the pressure of tourism does not represent much risk to the local environment. There is, of course, sewage, CO_2 and soot emissions from the power plant, and the carefully sorted fractions of litter, most of which is transported back to the mainland. Arctic tourism has existed for more than 150 years in Svalbard. It will continue doing so, but steps have been taken to limit the impact while maximising the benefit for the local community in the long run, to the extent that such a balance is possible. After visits in 1896 and 1897, the Swiss painter Hans Beat Wieland prophesised that Spitsbergen would become 'the tourist country of the future'. Clearly, the future had begun already by the end of the 19th century. ∎

Modern times III: research

An important pillar of today's economic activity in Svalbard is research. No place in the Arctic is so far north and so easily accessed as Svalbard. There are four research sites in the archipelago, which are manned all year round: Hornsund, Longyearbyen, Barentsburg, and Ny-Ålesund. Significant and unique Arctic research is carried out in Svalbard, which benefits from an exemplary infrastructure, such as access to airports, world-class laboratories, instrumentation, time series, and research ships.

Ny-Ålesund is the most northerly, permanent civil research station in the world (page 103). From its modest start as a coalmine and celebrated support base for many Arctic expeditions, it has developed into a modern base for international Arctic research and environmental monitoring. In 1964 the Norwegian Government decided that Ny-Ålesund would be the main site of Norwegian research in Svalbard. In 1966, Tromsø geophysical Observatory (also known in Norway as the Northern Lights Observatory) established a field station in Svalbard. The European Space Research Organization (ESRO) established a station in 1967 and in 1968 the Norwegian Polar Institute built the first research station to remain open all year round. Today, there are a number of Norwegian research stations in Svalbard, including the Norwegian Polar Institute's Sverdrup Station and Zeppelin Mountain Atmospheric Research Station, the Norwegian Space Centre (Norsk Romsenter), and the Kings Bay Marine Laboratory. Germany, China, France, India, Italy, Japan, South Korea, the Netherlands, and the UK all have permanent research stations in Ny-Ålesund. Approximately 20 nations have ongoing science projects and experiments in the settlement. The above-mentioned Zeppelin Mountain Atmospheric Research Station - located close to Ny-Ålesund, on the Zeppelin Mountain, which rises to 474 m above sea level - is one the most significant research sites in the entire Arctic. It is central for the monitoring of the global atmospheric environment. The data gathered there are of importance for the detection of climate change, changes in stratospheric ozone and UV, environmental contaminants, and long-distance transport of air pollution. The instruments's sensitivity levels are so high that the brief daily visits made by personnel can be immediately detected through changes in the quality of the surrounding air.

Longyearbyen is home to a most interesting institution, housed in an eye-catching building: The University Centre in Svalbard (UNIS) (pages 102 and 168). In this magnificent building, with its unique, 'organic' architecture that fits so well into the skyline of Adventfjorden, Norway offers Arctic studies to national and foreign students alike. Almost half of the students are foreigners and tuition is free – Norway's contribution to international understanding of the role of the Arctic in a global environment. Almost any field relevant to the Arctic, from technology to pollution and biology, is covered. The large number of students gives Longyearbyen a youthful feel. Among the various study branches, the one that has most relevance for Svalbard life is biology. UNIS has the only biology department in the European High Arctic, offering undergraduate education on a regular basis, graduate research experience, and postgraduate research. As the 'Gateway to the High Arctic' in the North Atlantic region, UNIS provides students and scientists with excellent academic facilities to study Arctic organisms and environments. Of special interest are adaptations of organisms to extreme physical conditions and the constraints to various biological interactions under these conditions. UNIS has active research programmes within both marine ecology and terrestrial biology/ecology, where functional mechanisms and processes in the Arctic are emphasised and co-operation across traditional disciplines is encouraged. UNIS covers four interrelated, main topics: biogeography, winter biology, sea ice, and population dynamics.

The Polish Polar Station is located at Isbjørnhamna, Hornsund fjord (see page 194). This research facility

operates continuously all year round and has been doing so for more than 30 years. The station is run by the Institute of Geophysics, Department of Polar and Marine Research, Polish Academy of Science, and carries out research in various branches of geophysics and the study of the polar environment. Excellent work has been carried out to describe and monitor the marine environment and a set of excellent biological keys was published (between 1990 and 1994).

The highlights of Polish marine research conducted in Svalbard are oceanographic work on the west Spitsbergen current to study climate variability and climate change, and the related time series generated from this work.

The waters engirdling Svalbard are the best-investigated marine regions in the entire Arctic and the ice-enforced research vessels – the *Helmer Hanssen* (previously the *Jan Mayen*, page 100) owned by the University of Tromsø, the *Lance* owned by the Norwegian Polar Institute, and the Polish vessel the *Oceania* – all play important roles. They spend many months a year around Svalbard and represent the backbone of marine research in the European sector of the Arctic Ocean. ■

Modern times IV: bustling urban life

inhabitants go to buy food and supplies. Upon entering the mall, one may ask: Is this place really in the High Arctic and a part of the fabled wilderness? One can wander among displays of tropical fruits, Mediterranean vegetables, fresh milk, the finest wines, high quality porcelain, and prestigious glass brands. Indeed, the quality of life is not merely excellent: it is breathtaking. It is, as so often in modern times, based upon goods transported from somewhere in the world, not sustained by local production. At times, one can be stunned by how artificial and externally controlled the world in Longyearbyen has been, and still is. Not only do the inhabitants live in this Arctic assembly with a lifeline to the mainland, the tourists that come by the thousands also contribute to a certain degree of artificiality. Longyearbyen has one of Norway's best restaurants. Rumour has it that one can also find the best wine cellar in Norway. There is a bar that has an unimaginable variety of beer from all over the world and bottles of Armagnac for each single year since World War II. The settlement is full of surprises.

Longyearbyen, like any other settlement in Svalbard, has never hosted a permanent, lifecycle society. No one has a permanent address in Svalbard. One cannot give birth to children, one cannot be buried there, there is no retirement home, and there is no upper secondary school or higher level of school education. People come and go, as has been the case from the very beginning. All is transient. In common with tourists, all 'visitors' to Svalbard have a time slot, and all know that it will come to an end: one has to return to where one came from. Living in Svalbard is a switch between phases of life on the mainland.

Unsurprisingly, modernity has also reached Svalbard in terms of internationalisation. Whereas the official language is Norwegian, many speak Russian or Ukrainian, and 12 of the total population of approximately 3000 are not Norwegians. Since 2002, Longyearbyen Community Council has had many municipal responsibilities, including utilities, education, cultural facilities, a fire department, roads, and ports. No care or nursing services are available, and no welfare payments are made. Norwegian residents retain their pension and medical rights through their respective mainland municipality. The hospital is part of University Hospital of Northern Norway, and the airport is currently operated by state-owned company Avinor. Ny-Ålesund and Barentsburg remain company towns with all infrastructures owned by Kings Bay and Arktikugol, respectively.

Longyearbyen is a bustling centre of activity in Svalbard. The main street is crowded with people at all times of the year. There are numerous shops of many kinds, but those selling sport equipment and clothes of the highest quality and design dominate. The main customers in these shops are seemingly the visitors, rather than the local population. There is also a shopping mall, the secret centre of the settlement, where all visitors and

In the safety and peace of Svalbard, an unlikely bouquet of people enjoy a unique lifestyle, living side by side. Tourists from many countries, Russians, Norwegians, visiting scientists, students, and various types of official visitors all mingle in the main street, the shops, and bars and restaurants. Their clothes vary from average, overly fashionable to those designed for rough outdoor life. Today's bustling life is also accompanied by several, large kindergartens that house more than 150 young children living in the settlement, thus reflecting the children-friendly policy of Norway. Longyearbyen has a lot of young life, as reflected in the children, students, young workers, and the remaining population that appears young at heart, regardless of the true age. All this is a far cry away from life in Longyearbyen in the first decades after World War II. Life in Svalbard is continuously subject to rapid changes, ranging from decade-long episodes to centennial periods. Svalbard will certainly change in the future, but the question is how? ■

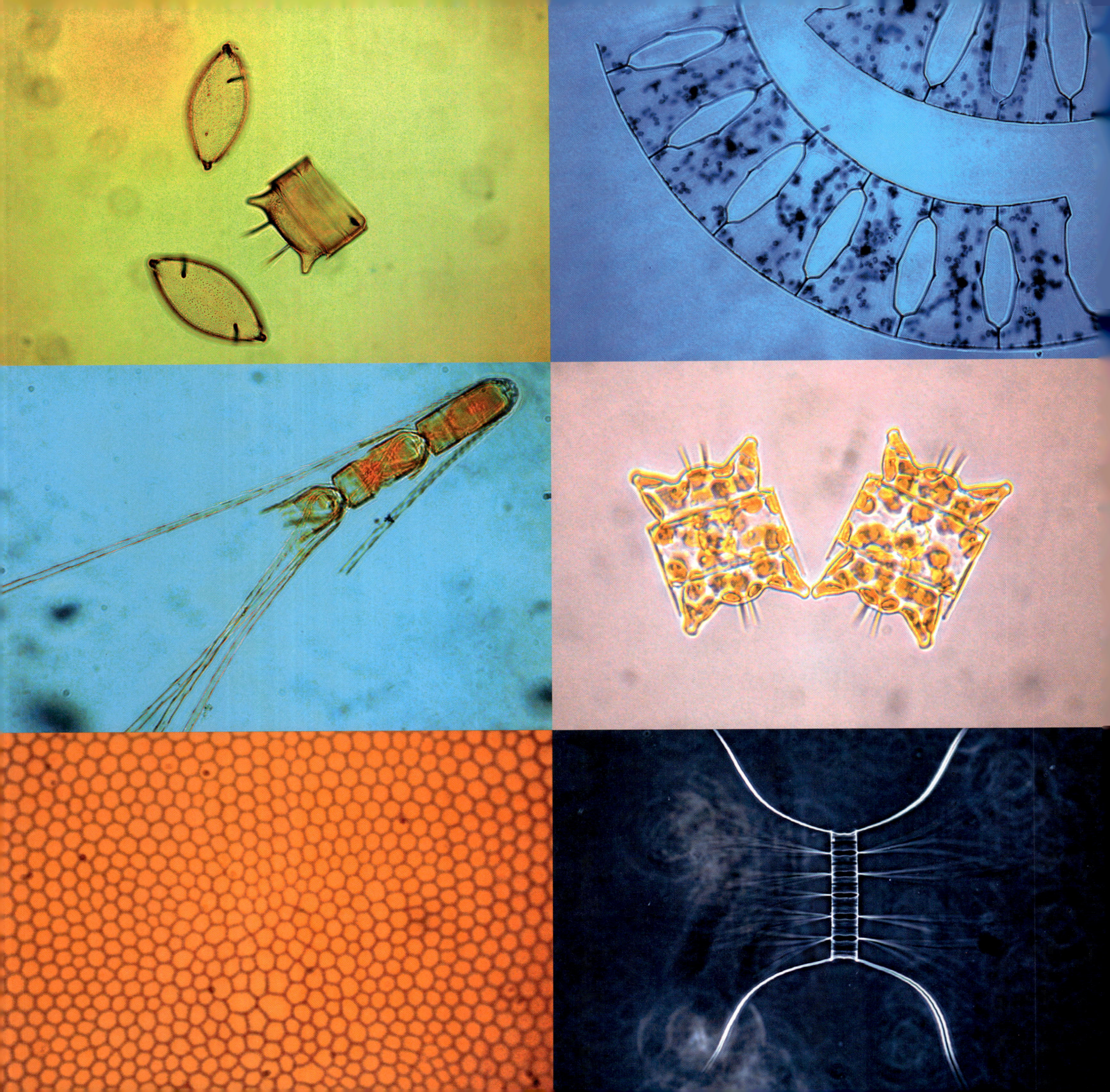

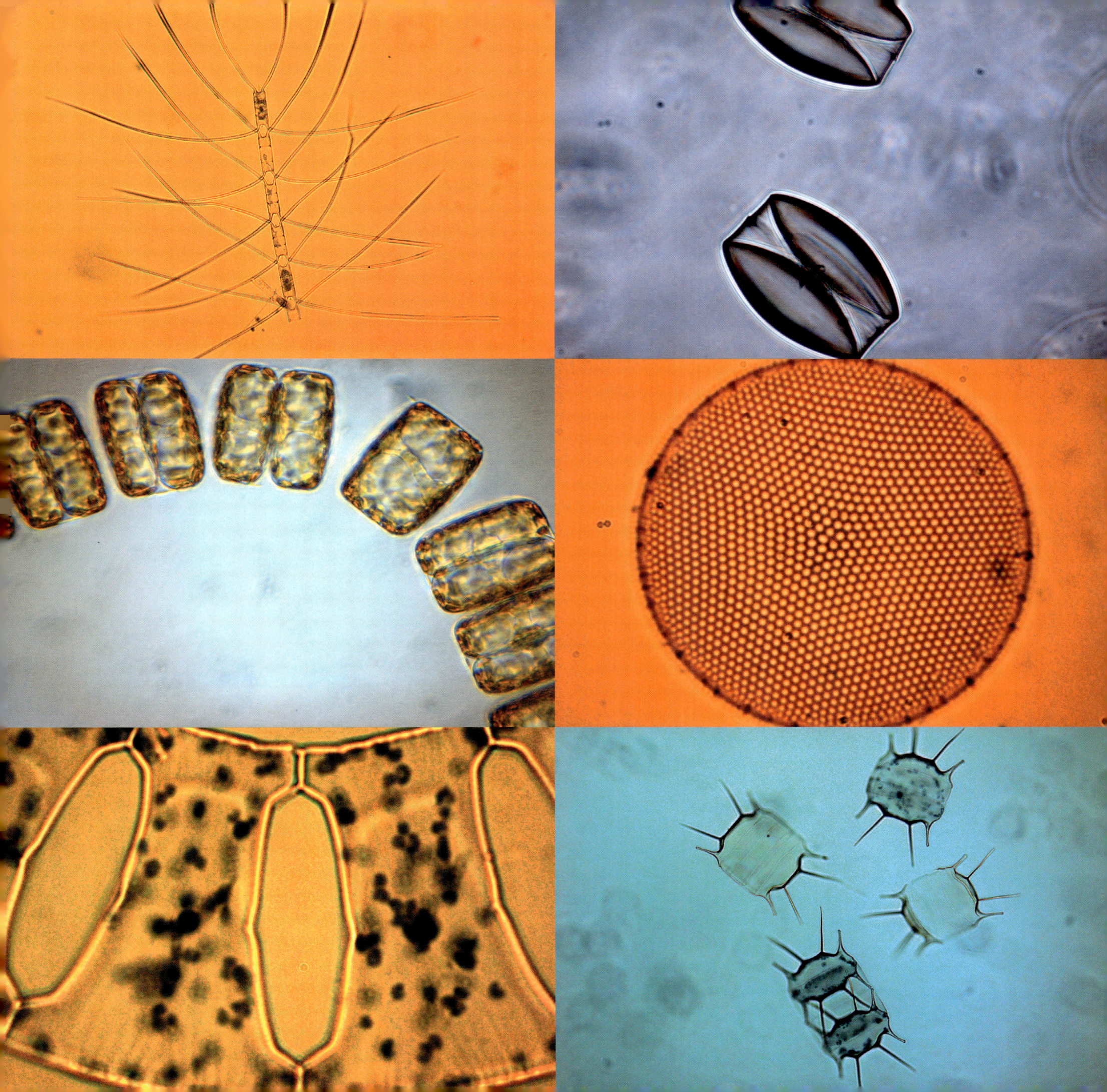

V. NATURE

PLANKTON
BENTHOS
FISH
MARINE MAMMALS
TERRESTRIAL MAMMALS
BIRDS

PLANKTON

What are plankton? The word 'plankton' is derived from the Greek word *planktos*, meaning 'wanderer' or 'drifter'. Plankton are any organisms that live in the water and are incapable of swimming against a current. They constitute a crucial source of food to many aquatic organisms, such as fish and whales. Plankton include plants, bacteria, and animals. Plankton are thus defined by their ecological niche rather than their developmental and evolutionary origin or taxonomic classification. Although most planktonic species are microscopic in size, plankton exist in a wide range of sizes, including large organisms such as jellyfish. In this chapter, I describe a few representatives of the various plankton species from Svalbard's marine waters and adjacent seas. The seas engirdling Svalbard, particularly to the south and south-east, are fertile and sustain one of the most important fisheries in the world. Compared to the abundant marine life, the terrestrial part of Svalbard should be characterised as a desert. Hence, I focus first upon the marine, 'real' life in Svalbard, starting with the most important segment of it, which is obscure to the average visitor: marine plankton.

The figure (right) derives from the book *Kunstformen der Natur* by the eminent German scientist Ernst Haeckel (1838–1919). It depicts planktonic radiolarians (0.1–0.2 millimetres), which are characterised by filigree mineral skeletons. ∎

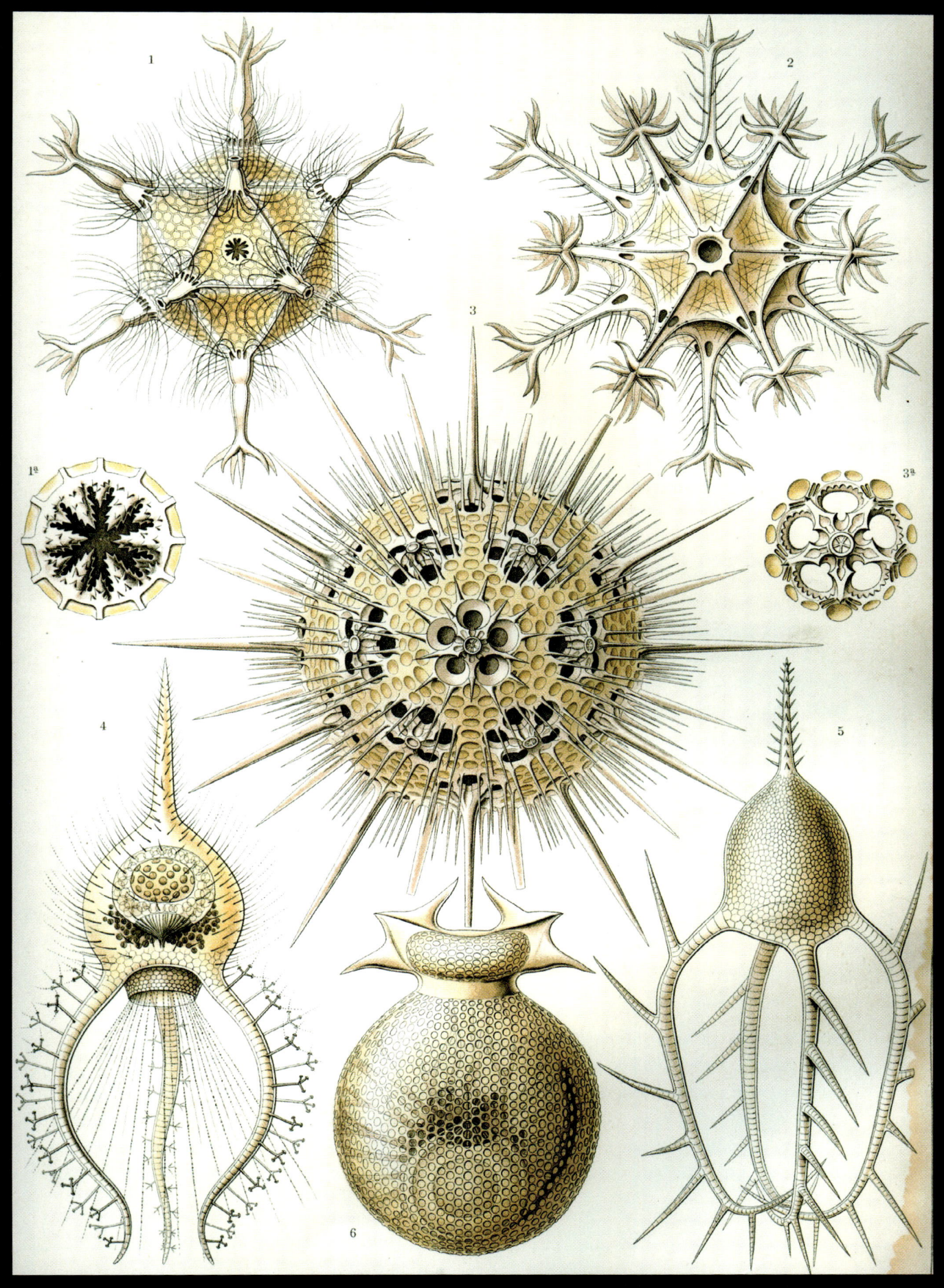

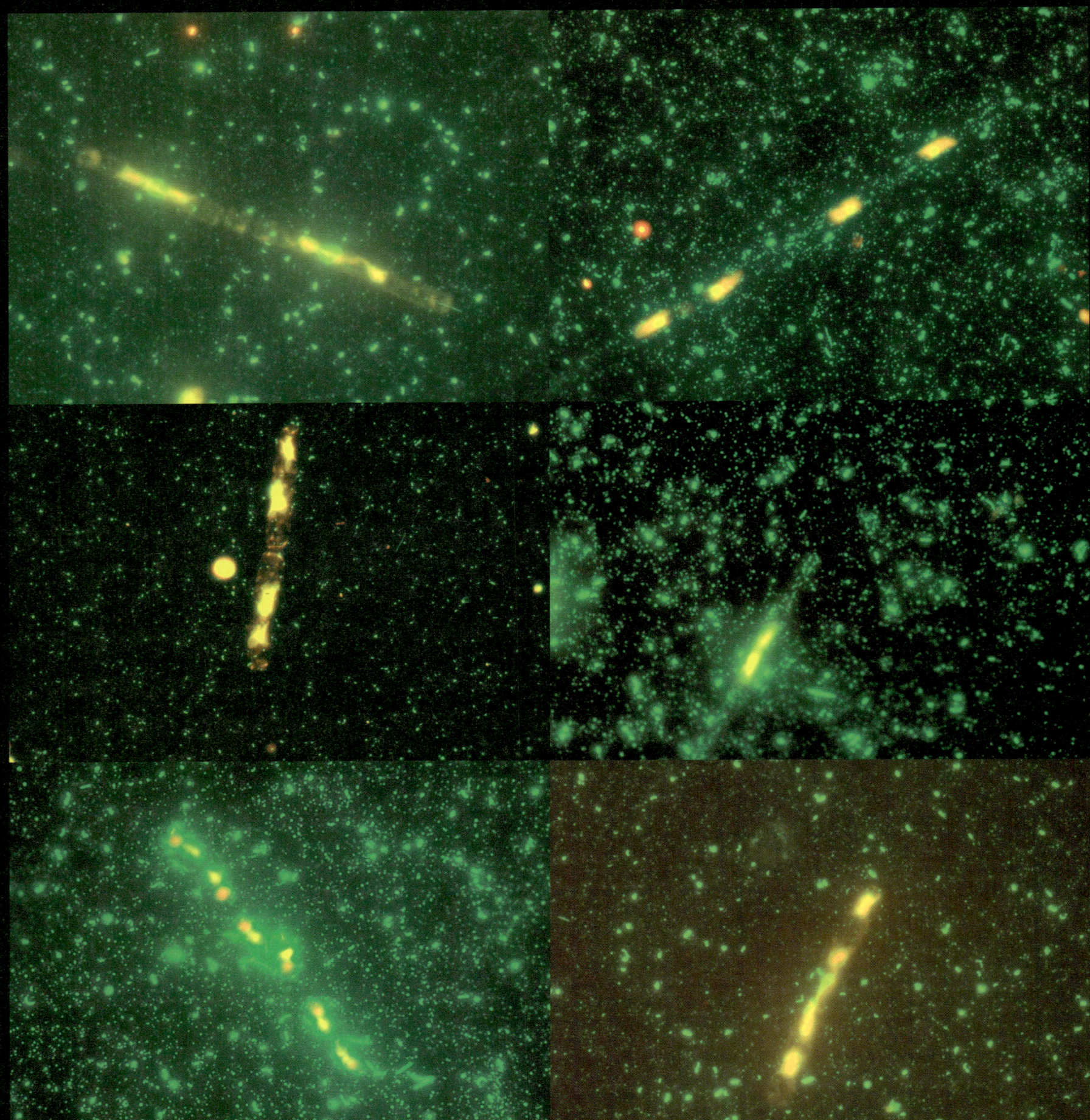

Plankton: marine bacteria

Many of us may think we know all there is to know about bacteria – they make us sick. However, in addition to plants, bacteria represent the very base of life on earth. None of the organisms that the majority of us are familiar with would be able to exist without them. This prompts the question: What are bacteria? Bacteria represent a large domain of single-celled organisms, typically a few micrometres in length. Bacteria have a wide range of shapes, ranging from spheres to rods and spirals. Bacteria are present in all habitats on earth, as well in organic matter as in the bodies of plants and animals. Bacteria are vital for recycling nutrients and hence for the productivity of the ocean. Contrary to popular belief, bacteria only rarely make us sick. Rather, life on earth rests upon the 'work' of bacteria.

Does the photography show bright stars of the universe? No, the 'stars' are fluorescent, marine bacteria visible under the microscope. The water sample was treated with synthetic colours, the magnification is ×4000, and the focus was extended to ensure sharp images of all cells. An ocean without bacteria would be a universe without stars. The oceans are brimming with more then 3×10^{28} bacteria. Can we imagine how much that is? Probably not, but it is about 100 million times more cells than there are stars in the visible universe. The total mass of bacteria in the ocean well exceeds the combined biomass of zooplankton and fish. The greatest living biomass on earth is thus one that we do not see at all. Most marine food web descriptions do not include bacteria or other microorganisms and very few include microbial links. Marine ecology is obviously facing a problem. Traditionally, marine ecology has focused mostly on larger organisms, preferentially those that man can harvest, such as fish, and less on the fundamentals of aquatic ecosystems.

The classic marine food chain – algae → zooplankton → fish – now has to be considered as a variable phenomenon in a sea of microbes. To obtain realistic perspectives, marine ecology has to master challenging times. When we visit Svalbard by ship and plough through the oceans, we should remember that the base of life in the ocean consists of plankton organisms that are very small and that among the smallest are bacteria. Thus, we sail through an ocean of bacteria and microorganisms that, in their perplexing minuteness, represent the foundation of life.

Plankton: diatoms

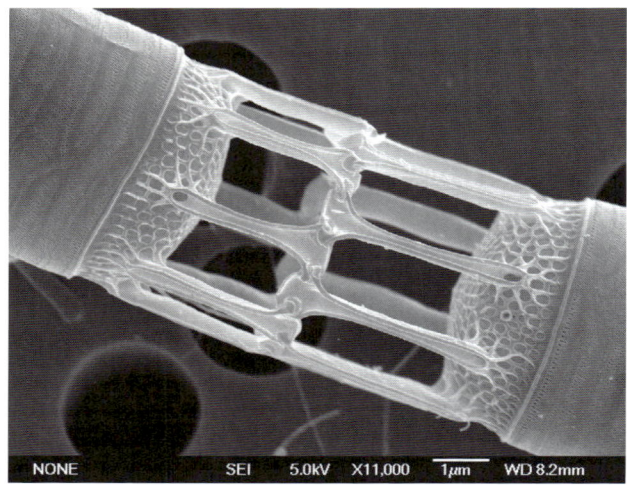

Diatoms belong to phytoplankton, i.e. the plant-type component of the plankton community. The name 'phytoplankton' originates from the Greek words *phyton*, meaning 'plant', and *planktos*. Most phytoplankton are too small to be seen individually with the naked eye. However, when present in high enough numbers, they may appear as a green discoloration of the water due to the presence of chlorophyll, the main plant pigment within their cells. Phytoplankton are photosynthesizing microscopic organisms that inhabit the upper sunlit layer of almost all oceans. They are agents for the creation of organic compounds from carbon dioxide dissolved in the water, a process that sustains the aquatic food web. Phytoplankton obtain energy through the process of photosynthesis and must therefore live in the well-lit surface layer of the ocean. Phytoplankton account for half of all oxygen present in Earth's atmosphere. Their cumulative energy fixation in carbon compounds is the basis for the vast majority of oceanic food webs. At the base of the oceanic food chains are important phytoplankton-feeding forms such as copepods, krill, and amphipods, which are food sources for fish and whales.

Phytoplankton comes in many shapes, size, and groups. They comprise an entire universe of plant types. There are about 5000 known species of marine phytoplankton. Among the common kinds are dinoflagellates (some of which produce toxins), green algae (often dominating in eutrophic regions), chalk-coated coccolithophores (which tint the water jade-green), and silica-encased diatoms. Here, I just refer to the latter group. Diatoms are unicellular, but can exist as colonies in the shape of filaments or ribbons. The silica cases occur in widely diverse forms, but usually consist of two asymmetrical sides with a split between them (see the stained examples on pages 106–107). Diatom communities are an important food source for copepods and krill. Usually microscopic, some species of diatoms can reach up to 2 millimetres in length.

In the waters around Svalbard, diatoms have an extremely important role for primary production (production of organic matter through algae as the first (primary) segment in the food chain; algae are the first segment because they have photosynthesis and thus they can form organic matter from carbon dioxide), particularly during spring and along the marginal ice zone. There are two different types of diatom blooms: those in open water and those at the bottom of or within sea ice. In both cases, the water or ice can be tinted green. Here, I show two pictures of diatoms. To the left is a scanning electron microscope image of the diatom species *Skeletonema marinoi*, which creates long strings formed of cells. Two of the cells are connected to each other. The magnification is ×11,000 and the bar scale is 1 micrometre. To the right is an underwater photograph of sea ice covered by variable amounts of ice algae. It is assumed that most of the green patches consist of diatoms. It is in such environments that copepods, krill, and amphipods can enjoy their first annual feeding orgies before the main phytoplankton bloom starts after the annual ice melt.

Plankton: copepods

Copepods are a small group of crustaceans. Crustaceans form a very large group of arthropods, including familiar animals such as crabs, lobsters, crayfish, shrimp, krill, and barnacles. The 67,000 described species range in size from 0.1 millimetres up to the Japanese spider crab, with a leg span of up to 3.8 metres. Crustaceans have an exoskeleton, which they moult in order to grow. In common with plankton, many copepods drift. Others live on the ocean floor or are parasites. Due to their huge numbers and biomass, and because water covers most of the planet, copepods are of great importance to Earth's ecology: they make up the largest single source of protein in the oceans.

Most planktonic copepods are about 1–2 millimetres in size as adults, although some grow larger. They are usually transparent and have an oval-shaped body with two large antennæ. Approximately half of the estimated 13,000 species of copepods are parasitic and have highly modified bodies. Free-living copepods mainly eat phytoplankton, and in turn serve as food for fish, mammals, and seabirds. Copepods do not digest all the algae they consume and consequently produce large quantities of sinking faecal pellets. Changes in copepod population composition and biomass may serve as an indicator of climate variability and change. Composition changes alter the variety of food available to fishes, mammals, and birds. Many species of copepods remain at different depths of the water column.

The Arctic waters species *Calanus glacialis* inhabits the shelves and edge of the seasonal ice zone, where they comprise the majority of zooplankton biomass. Their biomass increases as the ice recedes northwards in spring and summer. Global warming may force them to compete with their much-less nourishing, warmer-water relative *Calanus finmarchicus*, which spreads from the Norwegian Sea into the Barents Sea and around Svalbard. Warming may thus force the fat-rich *Calanus glacialis* into more northerly waters, which may have negative consequences for key fish species, whales, and birds.

Copepods contribute most to the animal production of the world's oceans. The surface layers of the oceans are believed to be the world's largest carbon sink, absorbing about 2 billion tons of carbon per year, which may be equivalent to one-third of human emissions. Many planktonic copepods feed near the surface at night and then sink (by changing oils into denser fats) into deeper water during the day to avoid being seen by predators.

Here, I show a side view of the Arctic copepod *Calanus glacialis*. The long spines on the antennae, the green food in the upper and lower part of the gut, a faecal pellet leaving the gut, the transparent lipid sack, and the tough upper exoskeleton are all clearly visible.

Plankton: krill

Krill is the common name given to shrimp-like marine crustaceans that are found in all oceans of the world. Many krill are filter feeders that use very fine combs to filter out their food from the water. Krill are considered an important connection at the bottom of the food chain because they feed on phytoplankton, zooplankton, or both. They convert these into a form suitable for many larger animals, for which krill makes up a large part of their diet. In the Southern Ocean, Antarctic krill constitute an estimated biomass that is roughly twice that of humans, yet in the Arctic their significance is far lower. Most krill species display large daily vertical migration, thus providing food for predators near the surface at night and in deeper waters during the day.

Around Svalbard, krill normally connect to warmer Atlantic waters and thus not so frequent in the north-eastern and eastern waters off Svalbard. Their remains are frequently found in the stomachs of fish and mammals. Around Svalbard, the majority of krill are omnivorous and only a few species are carnivorous. Northern krill have a relatively small filtering basket and actively hunt copepods and larger zooplankton. Krill are of less significance for predators than copepods in the Arctic. Nevertheless, krill are an important element of the marine aquatic food chain of Svalbard and adjacent seas.

The colourfulness of marine creatures is well worth seeing when they are alive! Accordingly, painters were invited to join early expeditions to paint or draw living organisms. An example of this approach, carried out in modern times by the Spanish zooplanktologist Miguel Alcaraz Medrano, is shown on page 108. The krill to the right shows the marvellous red pigmentation of the upper exoskeleton and the green pulp of consumed food and debris, just above the elaborate feeding basket with its numerous appendages. Also clearly visible are the three-dimensional facet eyes positioned at the top of stalks and which provide the creature with an unprecedented ability to see and detect.

Plankton: amphipods

Amphipods are crustaceans with a shrimp-like appearance with laterally compressed bodies and are generally known as 'side-swimmers'. The name 'amphipod' (differently footed) refers to the different shape of the appendages. Amphipods range in size from 1 millimetre to 340 millimetres. In the Arctic Ocean they generally range in length from 5 millimetres to 1.5 centimetres. Some species are red, green, or blue-green, but the majority are translucent, brown, or grey. The most well known species, which are very common in the littoral zone, are the sand-hoppers.

Amphipods exhibit a variety of lifestyles: they are crawlers, burrowers, swimmers, and tube dwellers. Some Arctic amphipods are common inhabitants of sea ice, where they feed on algae and smaller invertebrates.

Amphipods are often scavengers, scraping organic detritus (non-living particulate organic matter which typically includes the bodies or fragments of dead organisms, faecal material, and microorganisms) off the substrate, whereas others are herbivorous and filter-feed by extending their antennae into the water current as a net. Amphipods have separate males and females. Before mating, males often carry females beneath them. Sperm transfer occurs quickly, after which fertilised eggs are stored in a special chamber within the female. More than one brood per year is common. Amphipods play an important role as food for fish that hide beneath the ice. Sea ice is a demanding environment. It can freeze and produce highly saline fluids or it can melt and discharge freshwater. In addition, ice-melt implies that amphipods need to be able to survive without the protection of a complex and uneven substrate.

The creature shown here is neither a spaceship nor a monster from a fantasy novel. It is an ordinary amphipod from Svalbard's waters. Observe the tough upper exoskeleton, the highly developed, streamlined eyes, and the different appendages. What creature would stand a chance of survival when attacked by this armed, swimming fortress? ∎

BENTHOS

The term benthos comes from the Greek noun *benthos*, which means 'depth of the sea'. Benthos is the community of organisms that live on, in, or near the seabed. These flora and fauna live in or near marine bottom environments, from tidal pools along the foreshores, out to the continental shelf, and down to abyssal depths. At the fringe of deeper areas and with sufficient light, there is a large, complex, and bewildering variety of seaweeds – plants that are fixed to the bottom in the sea. Because light does not penetrate very deeply, the energy source for deep benthic ecosystems is organic matter from higher up in the water column that sinks. Such dead and decaying matter sustains the benthic food chain. Most organisms in the benthic zone are scavengers or detritivores, living on detritus or non-living particulate organic material.

The figure (right) is reproduced from the book *Kunstformen der Natur* by the eminent German scientists Ernst Haeckel (1838–1919). It depicts different types of sea anemones, arranged into a well-balanced illustration. ■

Benthos:
Marine seaweed

Seaweed is a loose colloquial term encompassing macroscopic benthic marine algae. Two specific environmental requirements dominate seaweed ecology: the presence of seawater and the presence of light in sufficient quantities to drive photosynthesis. The absence of sunlight in winter makes the Arctic Ocean an extremely harsh environment for plants. The sea ice and snow cover adds to the difficulties that seaweeds face, and only a few species can survive in Svalbard. Another common requirement is a firm attachment point. As a result, seaweeds most commonly inhabit the zone close to beaches, especially rocky shores in preference to sand or shingle.

When much of the Arctic ice melted during the last Ice Age, about 150 new seaweed species capable of living at low water temperatures and surviving prolonged periods of darkness claimed the Arctic seafloor. Most of these species grow under such cold conditions at similar or even higher rates than in warmer waters. The scouring action of sea ice prevents seaweeds and animals from establishing in the upper part of Svalbard shorelines. However, below a depth of 1–2 metres extensive 'woods' of seaweeds prevail, up to several metres high, as shown in the photograph. The impression of a barren, desert-like landscape – at best characterised as tundra – is in stark contrast to the rich vegetation below the surface of the water. The highest productivity levels in Svalbard are definitely found within the sea, not on land. Although Arctic seaweed serves mainly as refuge for underwater animals rather than food, during low tides it may be a source of food for land animals such as the Arctic fox.

The photographs show various aspects of marine seaweed ecology in Arctic waters: the mass of 'foliage' moving in currents and waves (page 124); sampling seaweed just above sea level (upper left); the surprising length of some seaweeds (lower left); and the 'leaf stalks' and 'leaves' of seaweeds undulating under a shimmering surface.

Benthos: hard bottom fauna

Marine hard bottoms feature some of the most spectacular and diverse biological communities on the planet. They not only contain a rich treasure of genetic, taxonomic, and functional information, but also deliver irreplaceable ecosystem services. At the same time, they are highly vulnerable and increasingly threatened by anthropogenic pressures.

Arctic hard-bottom substrates are not well studied, but a number of investigations have been carried out along the coast of Svalbard and within its fjords, including the longest time-series studies of benthos anywhere in the Arctic. Arctic glacial fjords contain a complexity of habitats influenced by strong environmental gradients. The best-studied region is the open fjord Kongsfjorden, which does not have a major sill. A mixture of Arctic and subarctic fauna and flora live there due to the influence of both Arctic and Atlantic waters. Through a cooperation between Norwegian and Russian taxonomy experts, an important baseline assessment has been made of the hard-bottom benthic species living in this fjord. The diversity and biomass of benthic hard bottom organisms is high in the middle and outer part of Kongsfjorden, and decreases toward the tidal glaciers of the inner bay. Species numbers (~300) show a similar pattern between the middle and outer fjord. In contrast, hard-bottom sites located in the inner bay contain fewer species and hence are considered less rich. Depth is the most important factor accounting for the varying density patterns of zoobenthos.

Climate warming will probably increase the boreal component of the fauna in Kongsfjorden, and northward expansion of distribution areas for boreal and more southern forms will probably lead to the introduction of new benthic species.

Sea anemones cover a rock onto which seaweeds have fastened their 'stalks'.

Benthos: soft bottom fauna

Soft-bottom habitats are unconsolidated, non-vegetated bottoms that occur in all marine systems. Soft bottoms have dynamic features, the distribution and character of which may change with shifting patterns of bottom erosion and deposition. Environmental characteristics, such as bottom grain size and distribution, salinity, dissolved oxygen, and flow conditions, will affect the condition of a soft-bottom habitat and the type of organisms that can be found there. The characteristic of soft bottom regions is the mobility of an unconsolidated, non-cemented bottom. Soft bottom acts as 'storage batteries' for nutrients and food particles, recycling them between the bottom and the water column, thus keeping the ecosystem in balance.

Clams and brittle stars dominate the surface of the soft bottoms of Svalbard fjords and continental shelves. Adult Arctic scallops can form dense colonies. They swim with a zigzag motion by clapping their valves together and forcing water through openings on either side of the hinge. Tiny benthic animals provide a vast food supply for both young and adult fish.

Many animals live hidden in soft-bottom substrates. Among them, worms and small crustaceans usually process great amounts of bottom material, either through the formation of burrows or by feeding. This results in continued turnover of mud and sand. Head-down, bottom-feeding worms can maintain a U-shaped burrow, the bottom of which may be many centimetres below the bottom surface. The worms feed upon bottom material, and pass faecal strands onto the bottom surface. Larger mussels on soft bottom are generally bottom feeders, taking up parts of the bottom by means of a mobile siphon, which extends to the surface. Soft bottom habitats provide a hiding place for burrowing marine animals such as clams and worms, as well as predators such as flatfish. Trawling is a problem for soft bottoms because it stirs up them up and removes surface organisms. Thereafter, it takes a long time for re-colonisation to occur.

The photograph depicts brittle stars that move over the surface of a muddy bottom, several hundred metres deep.

Benthos: Northern shrimp

The northern shrimp is highly appreciated by consumers and found in cold parts of both the Atlantic Ocean and Pacific Ocean. Many different common names are used, including pink shrimp, deep-water prawn, deep-sea prawn, great northern prawn, and crevette nordique. The northern shrimp lives at depths of 20–1300 metres, usually on soft muddy bottoms, in waters with a temperature in the range of 2–14°C. The distribution of the northern shrimp in the Atlantic ranges from New England and Canada's eastern seaboard, southern and eastern Greenland, Iceland, to Svalbard and the North Sea. In their 8-year lifespan, males can reach a length of 120 millimetres, while females can reach 165 millimetres. The northern shrimp starts out as a male, but after 1 or 2 years, their testicles turn to ovaries and they complete their lives as females.

Northern shrimps are an important food resource, and have been widely fished in Norway since the early 1900s, and later in other countries, following discoveries of how to locate and fish them. The shrimps are sold peeled, cooked, and frozen and are consumed as appetisers all over the world. They are a 'must' on an Arctic smorgasbord. Whenever one visits Svalbard by ship, one can do so in the knowledge that the largest and best shrimps are a long way down below, often in ice-covered waters. They are a hidden food marvel and probably the most widely spread Arctic food. Few connect this delicacy with a harsh and presumably hostile environment. ■

FISH

Arctic marine waters are home to about 240 species of marine and migratory fishes (sea to freshwater). The number of species may differ due to shifting of the Arctic faunal barrier over time, differences in taxonomic opinion, and the discovery of new species. Whereas the levels of productivity in the central Arctic Ocean are low, the subarctic regions such as the Bering Sea and the Barents Sea are highly productive and support some of the most valuable fisheries of the world's oceans.

Most Arctic Ocean marine fish species are benthic and live either on or closely associated with the bottom. A few fish species are pelagic, and move about freely in the water column. Other species are both benthic and pelagic, such as the ice-loving polar cod, which hides in crevices beneath the ice to avoid predators. Fish in the Arctic Ocean do not freeze because they have natural anti-freeze substances that keep their body fluids flowing at sub-zero temperatures. Cisco, whitefish, trout, char, and salmon all belong to the group of Arctic migratory fish.

A few of the most common fishes in Svalbard's waters are shown here, i.e. ones that readers may encounter, not necessarily during visits to Svalbard, but on their plate elsewhere. The photograph shows part of a trawler's catch, comprised mainly of cod, haddock (upper left), wolf fish (lower right), and red fish (top). ■

Svalbard Life | 133

Fish: Polar cod

The polar cod is related to the true cod and has a slender body, a deeply forked tail, projecting mouth, and a small whisker on its chin. It is silver in colour, with brownish spots. It grows to a length of 40 centimetres. The species is found further north than any other fish with a distribution mainly spanning the Arctic seas off northern Russia, Alaska, Canada, and Greenland. Polar cod is also found in the waters around Svalbard as well inside its fjords.

The fish lives for a few years and is most commonly found in surface waters, but is also known to descend to depths greater than 900 m. The polar cod is also known to frequent river mouths. It is a hardy fish that survives best at temperatures of 0–4°C, but may tolerate colder temperatures owing to the presence of antifreeze protein compounds in its blood. The fish group in large schools in ice-free waters or hide in cavities, crevices, or inside ridges under the ice.

Polar cod feed on zooplankton and krill. In turn, it is the primary food source of narwhals, belugas, ringed seals, and seabirds. Polar cod have been fished commercially in Russia. When sea ice is broken by a vessel and ice floes are turned upside down, every now and then a perplexed and dizzy polar cod will be carried to the surface and seabirds, such as kittiwakes, can stuff themselves with the fresh and excellent prey items. Many of the preferences and the life cycle of polar cod are still not well known. During the polar night, polar cod develop their gonads to prepare for forthcoming reproduction. They allocate enormous amounts of bodily resources to reproduction. Polar cod hold the world record for gonad development: before spawning, their gonads comprise 50% of their body weight!

Fish: Capelin

The capelin is a small forage fish found in both the Atlantic Ocean and Arctic Ocean. It is a key species of immense significance for the marine food webs around Svalbard. They constitute a rich food source for whales, seals, cod, and seabirds. Capelin make an annual feeding migration to the north, generally following the marginal ice zone. Northward-migrating capelin form a 'capelin front', which depletes their feeding grounds of available prey. Hence, they need to migrate further north continually. Subsequently, they migrate southwards to spawn off the coasts of northern Norway and the Kola Peninsula in north-west Russia. Capelin spawn on sandy beaches and sandy bottoms at the age of 2–6 years and have an extremely high mortality rate after spawning: the mortality rate for males is close to 100%. Males reach 20 centimetres in length, whereas females can grow up to 25 centimetres in length. They are olive-coloured on their back and silver on their sides.

The capelin is a specialised plankton feeder. It is the most important planktivorous fish and an ecological keystone species in the Barents Sea. Capelin graze on lipid-rich zooplankton and thus represent a crucial link between lower and higher pelagic trophic levels. The meaning of life everywhere in the Arctic is fat and capelin play an important role in this respect. Three groups of planktonic crustaceans dominate the diet of capelin and polar cod: copepods, krill, and amphipods. The northward-feeding migration pattern of capelin can be seen as an adaptation to exploit the plankton production associated with the receding ice edge. The value of migration is dependent on the biomass of available food as well as on the size of the capelin stock. It is suggested that about one-quarter of zooplankton production is suitable as prey for capelin. Capelin need to graze an area equivalent to more than half of the Barents Sea in order to sustain their stock levels by the end of each summer.

The size of the Barents capelin stock has varied widely in recent years. Based on indirect knowledge about stock dynamics, there are reasons to believe that fluctuations in capelin stocks are an inherent part of the ecology of Arctic seas. Possibly, they should be regarded more as natural perturbations than as reflecting man-induced instability in the ecosystem due to fishing. However, there is also evidence that the exploitation of capelin and its predator stocks may have affected the magnitude and length of periods when the capelin stock has been at low levels. The events observed have revealed complex and intricate relationships among ocean climate variability as a driving force and both zooplankton stocks and the growth rate and stock development of capelin.

Fish: Cod

Cod is a demersal fish species that has been and is of utmost significance for fisheries in the North Atlantic Ocean and the adjacent Norwegian and Barents Seas as well as for human consumption in these areas. Cod are popular as food, as they have a dense, flaky white flesh with a mild flavour. Cod livers are processed to make cod liver oil, an important source of various vitamins and omega-3 fatty acids. In Scandinavia and eastern Canada, cod fisheries have been essential throughout all times, both for export and consumption. The Basque and Portuguese specialised in fishing off Newfoundland and the outer Gulf of St. Lawrence to produce the beloved dish named bacalao (based upon salted cod dried on cliffs). Fresh cod is a traditional food primarily in Scandinavia, but also in regions adjacent to northern Europe. In the United Kingdom, cod is one of the most common ingredients of the traditional dish 'fish and chips'. Adult, spawning cod are caught to prepare stockfish (air-dried fish); particularly during medieval times, stockfish was one of the prime exports from Norway to Catholic Europe, where they were eaten during Lent. The trade was and is an essential corner stone of Norwegian economy.

Because cod is a top predator in the waters around Svalbard and Barents Sea area, its diet is a good indicator of the state of the ecosystem. Capelin is the most important prey for cod, whenever present, and other important prey are krill, polar cod, amphipods, and shrimps. There have been substantial changes in the fish populations of the Barents Sea during the past 30 years, caused by fisheries and natural variability. The depletion of capelin in the late 1980s clearly influenced cod stocks. A higher proportion of less digestible food with a lower calorific content resulted in cod that were reduced in length and weight, of increased age at sexual maturity, and had seriously reduced fitness. However, during a second capelin collapse (1993-1997), the negative effects on growth, sexual maturity, and the condition of cod were much reduced, possibly due to the increased presence of immature herring. Obviously, cod stocks – as with most other marine stocks – cannot be managed in isolation, but only in concert with other key organisms (food or predators).

The cod stocks in the Norwegian and Russian sectors along the coast of Svalbard and in the Barents Sea are probably among the best-managed stocks and fisheries in the world. Both during and since the Cold War, the guidelines for Norway and Russia have been co-operation between nations, the results of good quality research, and the will to act in a responsible manner by placing a higher value on sustainability and longer term economic outcome than on short-term profits. Both countries, and a wider world keen to consume high-quality fish, are satisfied with this rewarding approach. Tragically, too few seas have been impacted positively by such achievements. ■

MARINE MAMMALS

A total of 12 species of marine mammals regularly inhabit the Arctic Ocean: 4 species of whales, the polar bear, the walrus, and 6 species of ice-associated seals. Several additional species (e.g. the sperm whale, blue whale, fin whale, humpback whale, killer whale, and harbour porpoise) are spotted either occasionally or regularly within the marginal waters of the Arctic. Despite their distant relationships, all marine mammals share the common attribute of being predators that capture their prey on or in the oceans. All are relatively long lived, and virtually all are critical food resources to the indigenous peoples of the Arctic. This section presents a few of the more common species that may be encountered by visitors to Svalbard.

At times, whales become stranded and subsequently die in Svalbard. Such events create the biggest feasts in the Arctic, attracting polar bears, Arctic foxes and numerous gulls, all of which fight to secure a good meal. The photograph shows the remains of a whale that has been consumed by scavengers. During the climax of such feasts, polar bears literally throw themselves into the body cavities of dead whales, and their fur becomes tinted brown with coagulated and dried whale blood. ∎

Marine mammals: Polar bear

The Arctic animal par excellence is the polar bear, a bear that is largely native within the Arctic Circle encompassing the Arctic Ocean and its surrounding seas and landmasses. It is the world's largest land carnivore and the largest species of bear, in concert with the omnivorous Kodiak bear, which is approximately the same size. An adult male polar bear weighs around 350–680 kilograms, whereas an adult female is about half that size. Although the species is closely related to the brown bear, it has evolved to occupy a narrower ecological niche. Body characteristics are adapted for cold temperatures, for moving across snow, ice, and open water, and for hunting seals, which make up most of its diet. Although most polar bears are born on land, they spend most of their time at sea; the term 'maritime bear' derives from this fact. Thus, the polar bear is regarded as a marine mammal.

Polar bears hunt their preferred food of seals from the edge of sea ice, often living onshore on fat reserves when no sea ice is present. The preferred habitat is thus the seasonal sea ice covering the waters above the continental shelf and the Arctic waters between the islands of archipelagos. These areas, known as the 'Arctic ring of life', have high level of biological productivity in comparison to the deep waters of the High Arctic. This is particularly true of the seas surrounding Svalbard, which are among the most productive seas in the entire Arctic Ocean. Polar bears are therefore found primarily along the perimeter of the ice pack, rather than in the Polar Basin closer to the North Pole, where the density of seals is low.

The polar bear is classified as a vulnerable species, with several subpopulations in decline. For decades, large-scale hunting raised international concern for the future of the species, but populations rebounded after controls and quotas began to take effect. For thousands of years, the polar bear has been a key figure in the material, spiritual, and cultural life of several Arctic indigenous people.

Polar bears are found throughout Svalbard, including on the extensive sea ice cover south of the archipelago. Occasionally, they drift widely on sea ice, and there have been anecdotal sightings of polar bears as far south as the north Norwegian coast of Finnmark. It is difficult to estimate the global population of polar bears, but a working estimate is about 20,000 to 25,000.

Modern methods of tracking polar bear populations have been implemented only since the mid-1980s and are expensive to apply consistently over a large area. The most accurate counts require flying in a helicopter to detect polar bears, shooting tranquiliser darts to sedate them, and then following the tagged bears via satellite.

Young polar bears are curious and come often close to vessels (page 141). The photographs above shows a polar bear that has recently caught a seal, with gulls waiting for the leftovers.

Marine mammals: Ringed seal

Ringed seals are the most common and widely distributed seals in the Arctic, and can be found throughout the Northern Hemisphere's circumpolar oceans, where they feed on polar cod and a variety of planktonic crustaceans. Different populations have different names and exhibit some variation in their behaviour and appearance. Ringed seals derive their name from the light-coloured, circular patterns that appear on their darker grey backs. Some of these markings are so dense that they take on the appearance of splattered paint.

The ringed seal is the smallest and most common seal in the Arctic, with a small head, a short cat-like snout, and a plump body. Their coat is dark with silver rings on the back and sides, and they have a silver belly. Depending on subspecies and conditions, the size of adult ringed seals can be in the range of 100–175 centimetres. Ringed seals are commonly associated with ice floes and pack ice. The seals maintain a breathing hole in the ice, allowing them to use ice habitats that other seals cannot access. Their small front flippers have claws that are more than 2 centimetre thick and are used to keep the breathing holes open in up to 2 metres thick ice. Ringed seals prefer to rest on ice floes and in the summer they move farther north to denser ice. They can be found throughout the marginal ice zone around Svalbard, particularly in the south-western Barents Sea.

Female ringed seals reach sexual maturity at 4 years, whereas males do not reach maturity until they are 7 years old. The mating season starts in late April, when males roam the ice in search of a mate. Paired seals may spend several days together before mating, after which the male will search for another mate. During the spring breeding season, females construct dens within the thick ice, in which they give birth. After a gestation period of 9 months, they give birth to a single pup in March to April, on ice floes or shore-fast ice. The pups are weaned after 1 month, during which period they build up a thick layer of blubber. In that time, they are vulnerable to attacks by polar bears, which break through the roofs of the cavities in which the pups rest. Also during the pupping season, Arctic foxes and glaucous gulls take ringed seal pups born outside dens. Killer whales and occasionally walruses prey upon pups in the water.

Ringed seals are normally solitary animals and when hauled out on ice they keep a distance of hundreds of metres from each other. Feeding is usually also a solitary activity and their preferred prey includes shrimps, polar cod, and herring. When feeding, ringed seals may dive to depths of 10–45 metres. Ringed seals live for about 25–30 years

Marine mammals: Bearded seal

The bearded seal is a medium-sized seal species found within and near to the Arctic Ocean. Its common name is derived from its most characteristic feature, namely conspicuous and very abundant whiskers. When dry, the whiskers curl very elegantly, giving bearded seals a 'dashing' look. Adults are greyish-brown in colour, and darker on their back. Occasionally, their face and neck are reddish-brown. Bearded seal pups are born with a greyish-brown natal fur with scattered patches of white on their back and head. Bearded seals grow to about 2.1-2.7 metres in length and weigh in the range of 200-400 kilograms. Both sexes are about the same size.

Bearded seals prefer shallow Arctic waters less than 200 metres in depth. They also prefer areas with heavy ice floes or pack ice, where they can haul themselves out onto the surface. They generally segregate, with one adult per ice floe. Bearded seals on drifting ice floes can cover great distances, and their 'migration' is thus dependent on the season and distribution of ice floes. Bearded seals follow ice further south during the winter and further north during the summer. By riding drifting ice floes, they gain access to shallow waters, in which they feed. Bearded seals rarely choose to rest on land. However, in summertime, when ice floes are sparse, they have been known to haul out on land and gravel beaches.

Bearded seals are a primary food source for the Inuit of the Arctic coasts and for polar bears. The seals mainly find food at the bottom of the ocean, where they feed on a variety of small prey, including clams, squids, and fish; their whiskers serve as feelers in the soft-bottom sediments. Adult seals tend not to dive deep, but favour shallow coastal areas less than 200 metres in depth. However, pups up to 1 year in age will venture much deeper and dive down to depths of 450 metres.

Bearded seals give birth in the spring. Around Svalbard, seal pupping occurs in May. The pups are born on small, drifting ice floes in shallow waters and at birth they usually weigh in the range of 30-40 kilograms. They enter the water only hours after they are born, and quickly become proficient divers. The mothers care for their pups for 18-24 days, during which time the pups grow at an average rate of 3.3 kilograms per day. During that time, the pups consume an average of 8 litres of milk per day. By the time they are weaned, the pups have increased their weight to about 100 kilograms. If one imagines a weight gain of 3.3 kilograms based upon only 8 litres of milk it will quickly become apparent that the milk is actually liquid fat!

Just before the pups are weaned, a new mating cycle takes place. Female bearded seals ovulate at the end of their lactation period, but remain close to their pups, ready to defend them if necessary. During the mating season, male seals will 'sing', emitting a long-drawn-out call that ends in a sort of moan or sigh. This sound may attract females and/or may be used by the males to proclaim their territory or their readiness to breed. Bearded seals are promiscuous and have more than one mate during each breeding season. Males leave after mating and do not provide any care for their pups. Due to their solitary nature, bearded seals do not establish long-term bonds with mating partners.

Like many Arctic mammals, bearded seals employ a reproductive strategy known as delayed implantation. This means that the first stages of the embryo are not implanted for 2 months after fertilisation. Thus, a seal's total gestation period is around 11 months, although its active gestation period is only 9 months.

Marine mammals: Walrus

The walrus is a large marine mammal with a discontinuous distribution around the North Pole and subarctic seas.

Prominent tusks, whiskers, and bulkiness make adult walruses easily recognisable. Around Svalbard, adult males can weigh more than 1000 kilograms. The blubber layer beneath their extremely thick, dense skin can be up to 15 centimetres thick. Walruses live for about 20-30 years. The males reach sexual maturity as early as 7 years, but do not typically mate until they are fully developed, at around 15 years of age. Calves are born during the spring migration, from April to June. They weigh in the range of 45-75 kilograms at birth and are able to swim straight away. The mothers nurse for more than 1 year before weaning their offspring, but the young can spend between 3 and 5 years with their mothers. Young walruses are deep brown in colour and grow paler and more cinnamon-coloured as they age. Old males, in particular, become almost pink in colour. Walruses are social animals and live mostly in shallow waters above the continental shelves, spending significant amounts of their lives on sea ice. They have a diverse and opportunistic diet, feeding on a wide variety of marine organisms. However, they especially prefer clams, which they forage for by grazing along the sea bottom, using their tusks to disturb the layers in order find hidden food.

The walrus has played a prominent role in the cultures of the people of the Arctic, who have hunted the species for its meat, fat, skin, tusks, and bones. During the 18th and 19th centuries, walruses were widely hunted and killed in Svalbard for their ivory, skins, and meat. The Atlantic walrus was almost eradicated by commercial harvesting. Good estimates of walrus numbers are difficult to obtain, but the worldwide population is probably below 20,000. The population of walruses has rebounded somewhat since commercial hunting ceased. The populations in the North Atlantic and Laptev Sea remain fragmented and at low levels compared to the time before human interference.

Single or small groups of walruses can be seen in the northern regions of Svalbard, often near sea ice in relatively shallow waters, such as the protected, moraine-ridge island of Moffen, north of Svalbard.

Marine mammals: Minke whale

By the end of the 1930s, minke whales were targeted by the coastal whaling operations of many coastal states. By the early 1970s, following the overhunting of larger whales, minke whales became an even more attractive target. Hunting continued until the general moratorium on whaling was introduced in 1986. Although Norway initially supported the moratorium, an objection was placed with the International Whaling Commission. Norwegian commercial hunting was resumed in 1993. Currently, the country's annual quota ranges between a few hundred to just over 1000 whales, but often the quotas are not filled. The catch is exclusively for human consumption. Much of the annual quota is taken in late spring and early summer in waters around Svalbard. Those who wish to experience what life in the Arctic implies and do not have ethical impediments should use the opportunity to eat what has been one of the most common types of food for the Norwegian population: minke whale. Those that travel to Svalbard by ship should look out for whales that blow, particularly during 'whale weather', i.e. when the sea is comparatively smooth.

The minke whale or lesser rorqual (which takes its name from the French 'rorqual', which in turn derives from the Norwegian word *røyrkval*, meaning 'furrow whale') is the most abundant of all baleen whale species and is found in both the Arctic Ocean and the Antarctic Ocean. They are seasonal feeders, carnivores, and sieve their food through their baleen. They filter out plankton, krill, and small fish, and even chase schools of capelin. Minke whales are the second smallest of the baleen whales. Upon reaching sexual maturity at 6–8 years of age, both males and females measure about 7 metres in length. Estimates of maximum length vary from about 9 metres to 11 metres. Both sexes typically weigh 4–5 tons at maturity, and their maximum weight may be as much as 14 tons. Minke whales typically live for 30–50 years. In the north-eastern Atlantic, the total population of minke whales is estimated to be approximately 103,000. The gestation period for minke whales is 10 months, and calves measure in the range of 2.4–2.8 metres at birth. The newborn whales nurse for at least 5 months and in some cases up to 10 months. Calving is thought to occur every second year.

The minke whale has a black-grey colour and is distinguished from other whales by having a white band on each flipper. The upper part of the body is usually black or dark-grey and the lower part is white. Most of a minke whale's back, including its dorsal fin and blowhole, is visible when it surfaces to breathe. The whale breathes 3–5 times at short intervals, before 'deep-diving' for 2–20 minutes. The spout of the minke whale is comparatively low, almost an inconspicuous stream that rises to 2 metres above the water surface. A pronounced arching of the back precedes deep dives. The maximum swimming speed of minke whales has been estimated at 38 kilometres per hour.

Marine mammals: Fin whale

The fin whale is the second largest animal on earth after the blue whale, and can grow to more than 27 metres in length and weigh almost 74 tonnes. The long and slender body of the fin whale is brownish-grey with a paler underside. The fin whale is found in all of the world's major oceans, from polar seas to tropical waters, and often it can be sighted around Svalbard. It is absent only from waters close to the ice pack. Its food consists of small schooling fish (e.g. capelin), squid, and crustaceans, including copepods and krill.

The fin whale is usually distinguished by its tall, vertical spout, long back, prominent dorsal fin, and asymmetrical coloration. Full physical maturity is attained between the ages of 25 years and 30 years. Fin whales live up to about 100 years, although specimens have been found with an estimated age of 135–140 years. When feeding, they will blow 5–7 times in quick succession, but while travelling or resting they blow once every 1 or 2 minutes. On their terminal (last) dive before descending, they arch their back high out of the water. Thereafter, they dive to depths of up to 470 metres when feeding or only 100 metres when resting or travelling.

The fin whale is one of the fastest whales and can sustain speeds of 37 kilometres per hour, and bursts in excess of 40 kilometres per hour have been recorded, earning the fin whale the nickname 'the greyhound of the sea'. Fin whales are more gregarious than other rorquals, and often live in groups of 6–10 mammals. Female fin whales reach sexual maturity at 6–12 years of age, when their lengths reach 18–19 metres, and they reproduce every 2–3 years. Mating occurs in temperate, low-latitude seas during the winter, followed by a gestation period of 11–12 months. A newborn calf weans from its mother at 6–7 months of age, when it is 11–12 metres in length, and accompanies its mother to the summer feeding grounds.

The fin whale is a filter feeder, and feeds on small schooling fish, squid, and crustaceans. It feeds by opening its jaws while swimming at a relatively high speed (more than 10 kilometres per hour, which causes it to take in up to 70 cubic metres of water in one gulp. It then closes its jaws and pushes the water back out of its mouth through its baleen, which allows the water to leave but traps the prey. Each gulp provides a whale with approximately 10 kilograms of plankton. One whale can consume up to 1800 kilograms of food per day, leading scientists to conclude that the whale spends about 3 hours per day feeding to meet its energy requirements.

Like all other large whales, the fin whale was heavily hunted during the 20th century. The International Whaling Commission (IWC) has issued a moratorium on commercial hunting of this whale, although a few countries, including Iceland and Japan, have resumed hunting. Greenlanders also hunt the species under the Aboriginal Subsistence Whaling provisions of the IWC. Estimates suggest that the population of fin whales worldwide is around 100,000.

TERRESTRIAL MAMMALS

Landscapes in the High Arctic are determined by glaciation, permafrost and a harsh climate. Where permafrost is continuous, the ground is frozen to several hundred metres. Only in summer does the top 1 metre melt sufficiently to create poorly drained marshy soils with a certain amount of dry terrain. Arctic plants have to adapt to the harsh conditions of winter when there is intense frost, yet they also need to endure high summer temperatures that – within the low-lying vegetation – can rise to 30°C. Mosses and lichens are frequent. There is a scarcity of littoral plants, mainly due to sea-ice scouring. Inland, behind the shores, there is a profusion of special, small plants. There are some 165 species of highly adapted flowering plants in Svalbard. The growing season is very short, and therefore the plants need to grow, flower, and produce seeds within a period of just a few weeks. Despite difficult conditions, the flora can appear almost luxuriant during summer, with welcome splashes of colour that brighten the apparently barren landscapes. The purple patches of mountain saxifrage and the dark-green, almost fluorescent patches of vegetation below bird colonies (fertilised by guano) please the eyes of visitors that have started to become accustomed to the stony and icy deserts of Svalbard. Svalbard is an almost treeless region, but a handful of species, such as the Arctic willow and dwarf birch, survive by virtue of being small and stunted, often only ankle-high. A person from central Europe may find that the woods of northern Norway are mainly composed of bushes, whereas in Svalbard there are groups of shrubs.

A few of the most common mammals that a visitor to Svalbard may encounter are present here. The patterns in the photograph are the footprints of polar bears at the end of the snowy season. ■

Terrestrial mammals: Arctic fox

The Arctic fox is a small fox that is native to Arctic regions and is common throughout the Arctic tundra. The fox ranges between 70 centimetres and 110 centimetres in length. On average, males weigh 3.5 kilograms and females weigh 2.9 kilograms. The Arctic fox has a circumpolar range, meaning that it is found throughout the Arctic, including the periphery of Greenland, Russia, Canada, Alaska, and Svalbard, as well as in subarctic and alpine areas, such as Iceland and mainland alpine Scandinavia. The conservation status of the species is good, with the exception of the Scandinavian mainland population.

The Arctic fox lives in one of the planet's frigid zones. Among its adaptations for survival in the cold are its deep, thick fur, a system of counter-current heat exchange in the circulation in the paws, to retain core temperature, and a good supply of body fat. The Arctic fox has a low surface area to volume ratio, as is evident from its generally rounded body shape, short nose and legs, and short, thick ears. Since less of its surface area is exposed to the Arctic cold, less heat escapes the body. Its furry paws allow it to walk on ice in search of food. The Arctic fox has a keen sense of hearing and can precisely locate the position of its prey under the snow. When it finds prey, it pounces and punches through the snow to catch its victim. The fox's fur changes colour with the seasons: in the winter it is white, to blend in with snow, while in the summer it is brown.

Foxes tend to form monogamous pairs in the breeding season. Litters are born in the early summer and the parents raise the young in a large den. Such dens can be complex underground networks, housing many generations of foxes. Both parents help to raise their young. Young from a previous year's litter may stay with the parents to help rear younger siblings. The kits are initially brownish in colour and they turn white as they grow older.

In Svalbard, Arctic foxes generally eat any small animal, carcasses, or eggs that it can find. During April and May, they also prey on ringed sea pups when the young animals are confined to a snow cavity and relatively helpless. Fish beneath the ice can also be part of the foxes' diet. Arctic foxes also consume berries and seaweed, and are thus omnivores. If there is an overabundance of hunted food, the fox will bury what its family cannot eat. When its normal prey is scarce, the Arctic fox eat whatever they have buried earlier, scavenge leftovers and consume even faeces of larger predators, such as the polar bear.

Terrestrial mammals: Svalbard reindeer

The Svalbard reindeer is found only on Svalbard. It is very small compared to other subspecies of reindeer. It is approximately 150–160 centimetres in length, and weighs around 53–65 kilograms in spring and 70–90 kilograms in autumn. Svalbard reindeer may have a shoulder height as low as 80 centimetres. They are short-legged and have relatively short, round heads. Their coat is brown on the back and light on the belly. During winter, the coat is a lighter shade than during summer and often appears either light grey or yellow-white. The thick coat makes even starved animals appear fat in the winter and contributes to the short-legged appearance. The males develop heavy antlers during the period April–July and shed their velvet during the months of August and September. Males shed their antlers early in the winter. By contrast, the growth of female reindeers' antlers begins in June, and the antlers are usually shed after a full year.

Svalbard reindeer have a varied diet and will eat almost any type of vegetation. During winter, the reindeer concentrate on ridges, mountain slopes, plateaus, and other areas with comparatively less snow. During summer, they are found in areas where there is lush vegetation, often in valleys and on lowland plains. They spend most of their time feeding in the summer months and accumulate significant amounts of fat. The fat reserves are used during winter, when less vegetation is available and is of lower quality. Svalbard reindeer rarely occur in groups of more than 3–5 individuals, except during the rut in late autumn. The species has undergone several adaptations that help it to survive the variable climatic conditions and high degree of seasonality in Svalbard. For example, the animals exhibit very sedentary behaviour, which reduces their energy demands, and they have a well-developed ability to use their own reserves (both fat and muscle tissue) when access to food is very restricted during the winter. In addition, due to their fur, they are well insulated against the winter cold.

Svalbard reindeer mate in October. The males gather a harem of up to 10 females. The females are pregnant for about 7 months and give birth to a single calf in June. The calf suckles for about 3 months, growing rapidly during this time. Calves weigh about 3 kilograms at birth, but can gain 7–8 kilograms per month during their first summer. When conditions are favourable, females have their first calf at 2 years of age. The lifespan of Svalbard reindeer is normally about 10 years. The variable climatic conditions in Svalbard induce large amounts of variation in both survival rates and reproductive rates from year to year.

It takes a big stretch of imagination to link the Svalbard reindeer shown in the photographs to the fictional 'Rudolph the Red-Nosed Reindeer', with a glowing red nose, and his fellow chubby, short-legged reindeer pulling Santa's sleigh on Christmas Eve. ∎

BIRDS

The Arctic is seasonally home to some of the largest seabird populations in the world. More than 180 species of birds breed in the High North and take advantage of the short, but immensely productive summer season with an abundance of food. As the sunlight returns in spring, the burgeoning life at the ice edge, under the floes, and in the open waters of the ocean or polynias (areas of open water surrounded by sea ice) supports a huge influx of birds from the south. It has been estimated that some 16 million birds spend the summer in the Barents Sea segment of the Arctic Ocean alone. Because the Arctic winters are so cold, very few birds overwinter at high latitudes. When the summer arrives, the migrants turn up to exploit the summer food resources of insects, fresh underwater greens, and rich supplies of seafood. Because the birds need to be in the Arctic early enough to allow time for their eggs to hatch when the food bonanza starts, they lay on reserves of fat to see them through the lean times.

Here, I present a few of the most common species that a visitor may see in Svalbard. From the photograph, one can gain an impression on the environments in which kittiwakes raise their chicks. ■

Birds: Little auk or dovekie

The little auk is the smallest of the European auks, which constitute the most abundant bird species in Svalbard. It is probably one of the most abundant seabird species, has a circumpolar distribution and very rough estimates suggest a breeding population of some 37 million pairs worldwide. Little auks breed in large colonies, mainly on coastal mountainsides. The largest colonies in Svalbard are in the south-west and north-east, close to productive waters. Among these, the colonies in the fjords Magdalenefjord and Hornsund are considered the largest of all colonies in Svalbard. The colony in Hornsund has been investigated by Polish scientists for many years, who have studied, for example, what little auks feed upon, how far they fly in search of food, and their survival rates. Little auks nest beneath large rocks and lay only 1 egg per clutch. Both parents take an equal part in the incubation of the egg and subsequently the food search for the chick. Little auks mainly feed on crustaceans and their preferred prey is the Arctic copepod *Calanus glacialis*. Compared to its sibling species associated with Atlantic Water, the latter Arctic copepod stores comparatively large amounts of lipids. This characteristic is an adaptation to a life in the Arctic when there are lengthy periods with food shortages. The lipid-rich copepod is perfect food for little auk chicks, which have to grow fast in order to be ready to leave the nest before the approach of the next winter. If the preferred food is not avalaible in the vicinity of the breeding colony, the parent little auks either have to fly longer distances in search of it or switch to less energy-rich food. Either solution will affect the growth and survival of their chick. Studies have shown that the little auks can fly up to 300 kilometres in search of energy-rich food.

Birds: Arctic tern

The Arctic tern is a small, slender, white seabird that has a circumpolar breeding distribution covering Arctic and subarctic regions. The species is the only tern that breeds in Svalbard and has the longest regular migration route of any living animal – an extraordinary pole-to-pole migration, which means that the bird sees two summers per year. This small seabird breeds throughout the coastlines and tundra of Svalbard. Recent studies using tiny tracking devices have shown that Arctic terns can fly more than 70,000 kilometres per year, equivalent to two times around the world. Considering that this bird is long-lived and can reach 30 years of age, potentially they may fly a distance corresponding to 2–3 round trips to the moon during their lifetime. Arctic terns breed in colonies and are very aggressive when it comes to defending their nests and young from potential predators, such as Arctic foxes, seagulls or visitors. To avoid predation, the nest and chicks are very well camouflaged. Males and females are similar and cannot be distinguished in the field. They usually mate for life and often return to the same colony year after year. The female usually lays 2–3 eggs, and both parents share the task of incubation. Arctic terns mainly feed on small fishes, such as polar cod, and marine invertebrates, which they catch from the surface of the water or by performing shallow dives. The breeding population of Arctic terns in Svalbard is difficult to assess, but probably less than 10,000 pairs. The breeding population is considered stable.

Birds: Glaucous gull

The glaucous gull is the most abundant of the large seagulls in the Arctic. It is a circumpolar High Arctic species that commonly breeds in Svalbard. The glaucous gull is a large, pale gull with a large bill, and can be distinguished from the herring gull by the lack of black tip on their wings. It is also significantly larger than the more slender Icelandic gull, which otherwise can appear similar. Like most other species of large gulls, the glaucous gull is a '4-year gull', meaning that it takes 4 years to acquire adult plumage. The breeding population in Svalbard is estimated to be 4,000–10,000 pairs. Males and females are similar, but the males are slightly larger than the females. Glaucous gulls and great skuas are the only predatory birds of importance breeding in Svalbard. Their position at the top of the food chain make them vulnerable to the accumulation of various pollutants. Researchers from the Norwegian Polar Institute and Akvaplan-niva in Tromsø have studied the impact of pollutants on glaucous gulls breeding on the island of Bjørnøya. As the seagulls become old, concentrations of pollutants can be very high and even deadly, particularly in old male birds. Most glaucous gulls spend their entire life in the High Arctic region. However, a few can sometimes be seen along the Norwegian coast as far south as the North Sea, usually after periods of strong northerly winds, particularly during winter, or possibly when birds have followed fishing vessels returning south with their catch.

Birds: Northern fulmar

The northern fulmar is the only storm petrel to breed in Svalbard. The birds occur in various colour morphs, from white-light grey to relatively dark grey. The dark grey morph dominates in the north, including Svalbard, whereas the light morph is more common in the southern part of the species distribution range. Anyone who has been on a boat trip off the coast of Svalbard probably has seen this bird. It is highly abundant and often follows ships. Northern fulmars can appear similar to seagulls but are not closely related to them. Rather, they are petrels and their bills have tube-like nostrils to excrete salt. They cover huge distances in their search for food, which mainly comprises of various crustaceans, winged snails, squid, and small fish caught on or near the surface of the water. Northern fulmars spend all their time at sea and only come close to the shore during the breeding season. They are long-lived and have a low reproductive rate. They first reach sexual maturity at the age of around 10 years and lay only 1 egg per clutch. The chicks are fed by both parents and defend themselves by spitting an oily gastric juice at intruders. This juice is very unpleasant to those subject to it and it is difficult for predatory birds to remove it from their feathers. Consequently, as such predators become aware of the chicks' tactics they become more careful to avoid them. Northern fulmars can grow very old and recordings of birds ringed in Great Britain prove that they can become more than 50 years old. Each couple pairs for life. In Svalbard, the largest number of breeding pairs is currently found on Bjørnøya. The population size is difficult to estimate, but the breeding population is believed to be stable.

Birds: Snow bunting

The snow bunting has the most northerly breeding range of any songbird in the world, and the only passerine that commonly breeds in Svalbard. Snow buntings breed throughout most of the archipelago, but their population size is not known. The first birds, adult males, arrive in early April, when the ground is still covered with snow, whereas the females and young birds arrive a few weeks later. The males establish themselves, protect their territory, and sing beautifully to attract females. Many tourists visiting Svalbard during summer are surprised to hear such melodious birdsong so far north. The snow bunting mainly feeds on seeds, but also catches insects, particularly to feed its chicks. The female builds a nest of various plant material as well as feathers and hair. She incubates 4–7 eggs for about 2 weeks before they hatch. The male feeds the female while she is incubating, and later on both parents share responsibility for feeding the chicks for 3 or 4 weeks. The snow bunting is a migratory bird, leaving Svalbard in autumn to migrate south and east towards the Russian steppes north of the Caspian Sea and to Kazakhstan. The migration route from Svalbard to the overwintering area is not well known. New technology involving the development of small lightweight tracking devices hopefully will shed light on the journey taken every spring and autumn by the small songbird.

Birds: Kittiwake

Among the kittiwakes only the black-legged kittiwake is found in Svalbard. It is a middle-sized gull with black wing tips and a flight pattern that resembles more closely that of a tern than other gulls. It is the most numerous gull species in the world and lives most of its life at sea. Kittiwakes have a circumpolar distribution and they breed in the boreal and Arctic zones of the northern hemisphere. Kittiwakes are among the group of birds that regularly breeds in Svalbard and can be found along the coast, in open water, and on ice-covered waters throughout the entire archipelago. More than 200 colonies of variable size have been registered, ranging from a few birds to more than 100,000, totalling of 270,000 thousand breeding pairs. Currently, the largest colonies are found on the islands Bjørnøya and Hopen. The estimated numbers of breeding pairs in the Barents Sea region and Europe as a whole are about 900,000 and 2.1 million respectively. Kittiwake colonies are extremely boisterous and a cacophony of 'kittivaik, kittivaik' sounds fill the air. Black-legged kittiwakes overwinter throughout the North Atlantic and are not classed as migratory birds. Some kittiwakes arrive in Svalbard as early as February, but the majority arrive in March and April. In September, they leave Svalbard and silence rules again over the colonies.

The birds take their food from the sea surface, often preying as a flock. Kittiwakes are numerous in ice-covered waters and at glacier fronts, where upwelling freshwater brings food to the surface. They feed on invertebrates and fish at depths of up to 15–20 centimetres; capelin, polar cod, and amphipods are their common prey in Svalbard. Kittiwakes also prey on scrap fish thrown overboard from trawlers and their diet changes according to region and season. Various birds, such as glaucous gulls, feed on kittiwake eggs and chicks. The nests are built on narrow shelves and small projections of mountainous cliffs, a few to several hundred metres above sea level (see page 154-155). Frequently, visitors to Svalbard see kittiwakes using human constructions for their nests, such as window frames or other protrusions in Barentsburg and Pyramiden. Kittiwakes usually lay 2 eggs per clutch at the beginning of June, and the chicks hatch after 25–32 days of incubation by both parents. Kittiwake chicks are fed regurgitated, half-digested food by both parents and are able to fly after 5 or 6 weeks. After 2 years at sea, the young birds return to the colony where they were hatched. They become sexually mature after 3–5 years and the maximum registered age of a kittiwake in Svalbard is 19 years.

VI. COMMUNITIES: THEN AND NOW

LONGYEARBYEN
PYRAMIDEN
BARENTSBURG
SMEERENBURG
VIRGOHAMNA
NY-ÅLESUND
NY-LONDON
SVEA
HORNSUND
HOPEN
BJØRNØYA

Longyearbyen

Longyearbyen is situated in the small Adventfjorden, a side arm of Isfjorden (page 34), and is the main Norwegian settlement in Svalbard, with approximately 2500 inhabitants. The township functions like a capital, but Svalbard is not a country. The administration, the governor, the main airport, and most of the public services of Svalbard are located in Longyearbyen. First called Advent City, the settlement was renamed Longyearbyen in 1926 and thereafter remained the focal point of Norwegian interests in Svalbard. The Norwegian state-owned coalmining company Store Norske Spitsbergen Kullkompani (SNSK) played a major role in the development of Longyearbyen.

After World War II, only parts of the previous settlement were rebuilt close to the fjord. Meanwhile, a new settlement named Nybyen was developed within the valley, close to the mines that were in operation at the time. It was a time when Svalbard was often isolated because ship traffic was blocked by ice for many months each year. With the decline of coalmining and a changing world economy, the Longyearbyen centre developed closer to the fjord, to its present position. The opening of the airport and the end of the Cold War changed Longyearbyen. Today, it is a bustling village with colourful, modern houses, good roads, a wide range of shops, hotels, restaurants and coffee shops, a bank, a hospital, kindergartens and schools, a culture house, a library, and last, but not least, The University Centre in Svalbard.

Upper left: The transportation of coal at the beginning of coalmining in Longyearbyen in 1907, painted by L. Tinayre (see page 260). The painting is on display at the Musée Océanographique de Monaco. Right: Longyearbyen as it is today. Remaining towers related to coal transportation can be seen, as well as the river gravel banks and the pipeline (on poles) conducting heat, a characteristic permafrost construction.

It is a special experience to visit or live in Longyearbyen. The locals are far more open and relaxed than in most other places on Earth. The access to consumer goods is breathtaking given that one is surrounded by wilderness. There is a bustling outdoor life and there are hordes of visitors and tourists. Many foreigners live in Longyearbyen, among those a substantial Thai population. In addition, there are recurrent visits by people from the Russian settlement of Barentsburg. All this characterises Longyearbyen as having an international, extrovert, open, and energetic 'outdoor' lifestyle.

There was a strict separation between SNSK workers and administrative personnel at Longyearbyen far into the 1970s. The living quarters, food, and access to alcohol were quite different and the lines of separation between the two groups were significant. The hotel called Funken (from Norwegian funksjonær = salaried employee) accommodated administration, elevated from the valley bottom, with a view, and exposed to early sunshine, while workers were fed in the biggest building at the bottom of the valley, Huset (the house), which did not receive so much sunlight. The connections between the various settlements are also noteworthy. During the Cold War there were few if any connections between the two populations, except for consultations at administrative level, which were diplomatic in character. There were strong Soviet protests against the building of the Longyearbyen airport. The Soviet helicopter base at Cap Heer, which is far larger than there is need for, gave rise to Norwegian concern about Soviet plans. These are two of many examples illustrating that both parties were on constant alert. Svalbard was situated in the middle of potentially ballistic battlefields between the USSR and USA/NATO. Thus, both Norway and the USSR were keen to ensure that not the slightest indication of militarisation took place. These times are over and relations in Svalbard today are open and friendly!

Large windows of the modern building of the University Centre in Svalbard reflect the adjacent landscape and buildings (page 168). Upper right: two photographs of the mine and shoreline structures in 1907. Lower right: remnants of buildings that were destroyed during WWII. ∎

Svalbard Life | 169

Pyramiden

Inside the majestic Billefjorden, part of Isfjorden, lies the abandoned Russian settlement of Pyramiden (page 34). The name derives from the shape of the spectacular mountaintop behind the village, which resembles a pyramid. A Swede took possession of the Pyramiden coalfield, but mining success was limited, and in 1926 the property was sold to Russia. In 1931 the mine was taken over by the mighty Russian state-owned company Trust Arktikugol, which also is in charge of the remaining Russian interests in Svalbard. Pyramiden was evacuated in 1941, but mining started up again in 1946. In its heyday, approximately 700 men, 250 women, and 70 children lived in Pyramiden.

Pyramiden comprises all the elements that one connects with a modern, socialistic settlement. The dominating buildings are the culture and sport buildings, the four-floored living quarters, and the hospital. There was a kindergarten, a school, a swimming hall, central heating, stables for cows, indoor sties for pigs, and a greenhouse. The buildings are placed along a thoroughfare of boulevard dimensions, overlooked by a bust of Lenin, who, petrified by today's chaotic world, keeps this dream of a socialist city under surveillance. The soil for the grass banks and even the grass along the boulevard were transported from Russia to Pyramiden, paying tribute to an archaic, Russian idea: the holy earth of Mother Russia that supported Russians throughout the ages. The mountain, with its coalmine behind the village, the boulevard, and the majestic Nordenskiöldbreen calving into Billefjorden opposite, create a grand, theatrical setting. Gorbachov's glasnost provided a blow to Pyramiden's splendid isolation: a hotel for tourists was opened in 1990.

The settlement was abandoned suddenly in 1998, and the entire population was given just a few days' notice to leave. What they took with them were a few personal belongings. All the equipment was left behind, because it was not owned by any of the inhabitants, most of which were from Ukrainian mining districts. They had felt detached and were just working there, whereas all the rest was owned and run by Trust Arktikugol, on behalf of the Russian state. Thus, a well-equipped ghost town was left in a remote section of an Arctic fjord. Frost action, river displacement, regular theft and vandalism have all damaged the settlement over the years, but now the Norwegian state, through the governor of Svalbard, has protected the complex. Trust Arktikugol has been slow in meeting obligations to clean the place of waste, but progress has been made.

The hotel is once again open, for parts of the year, although few tourists seem to stay there. With the exception of one or two Russian visitors, all of the streets are empty. Surprisingly, however, in the middle of shabby building complexes there is a bar. It is open in the summertime, when, as a cosy human oasis, it appears like a mirage. It has impeccably clean wooden furniture and wall panelling, and, as shown in the image on page 206, the staff behind the counter wear tall, white caps and smile their best and most polite smile. How many visitors come to an unexpected bar in an unlikely location? Visitors are touched by the will and motivation of humans that create gentle retreats, under the totally unexpected circumstances of a petrified, socialistic dream city. Whereas the buildings outside cast shadows of the past onto the present, individuals inside create their own personal niches. Although changing political ideas and economic conditions create unpredictable conditions, humans shape their own realties, against all odds. Today, not even Lenin's stern glance (page 164) can prevent people from creating their own life in Svalbard. Neglect and suppression of the individual ultimately led to the end of communism.

It is indeed fascinating to visit the combination of a Soviet model city, of a ghost town, and the mesmerizing Arctic setting of Pyramiden. One can sense the aspirations of the people who went so far north to live there, raise their families, and had hopes and desires for a better future. Shortly after 1998, one could have witnessed a fully equipped small hospital, children playgrounds, an impressive library, music instruments, and sport equipment, all in the middle of nowhere and void of the essential ingredient of every settlement, namely inhabitants. The signs of people having left in a hurry are everywhere. Pyramiden was a perfect dream of a socialistic city that surprisingly enough is not situated in Russia, but in the capitalistic west, the High Arctic, and at a location that is as remote as possible. Utopias are frequent and represent some of the most dedicated wishes of humans to create new and better worlds. It is moving that the idea of a well-functioning society has resulted in such an example, where the will to change and improve the world gave rise to such a remote settlement.

Left: Russian children lost a toy wrench during play.
Right: the welcoming sculpture of Pyramiden close to the harbour, and the view from a building (inhabited by kittiwakes) on the glacier Nordenskiöldbreen.

At the start of the boulevard, or *prospekt* as it is called in Russian, stands a sculpture that underlines the dominating role of Trust Arktikugol. A red star hovers over a blue globe. A depiction of a tousled, dog-like polar bear resides on top in a red flag. Mosaics on the walls of the main building at the end of the boulevard express dreams of a brilliant and radiant future through the clichés of socialism: happy workers, helicopters, and atomic power plants. Humans and machines belong to the core ideas of the socialistic world. With so much dominating nature around, a far cry away from the technology dreams reflected in the images on the wall, one wonders how realistic this utopian, socialistic avant-gardism is.

Pyramiden is a utopia nourished by extreme visions and placed in the extreme north. It has become a relict and museum of Soviet revolutionary constructivism, on Norwegian soil. It is a Russian 'yes, we can' statement, that, over time, has turned into 'yes, we could'. The Norwegian author Kjartan Fløgstad describes it as an idea emptied of content, frozen in time, tormented by Arctic frost, by economic market trends, the Cold War, and a victorious capitalism. It is ironic that revolutionaries that successfully reform history turn into totalitarian rulers, whether pharaohs or comrade leaders of the Soviet Union. The final conclusion reached in Fløgstad's book on Pyramiden is that 'The mausoleum over one of these rulers is left empty in Svalbard. Pyramiden is the reversed grave that does not point down into the ground, but to the sky, where all see each other on TV all the time'.

A naked doll, spiked to a window, overlooks this utopia and wonders what has happened to the inhabitants' and planners' aspirations for a good, rich, and prosperous life? A childhood dream and the naivety that the world becomes a perfect place through an anti-individualistic socialism alone, appears to have terminated suddenly.

Barentsburg

The only active Russian settlement in Svalbard at present is Barentsburg, situated in Grønfjorden, a side fjord of Isfjorden (page 34). Russia has been the only country that, apart from Norway, has made substantial use of the terms and conditions of the Svalbard Treaty. As already mentioned, during the Cold War, relationships between the Norwegian and Russian settlements were quite reserved, but friendly. Nevertheless, the short distance between Longyearbyen and Barentsburg resulted in generic suspicion and both parties kept a watchful eye on each other. Russia will surely also in the future be a key player in the development of Svalbard.

The main commercial activity close to Barentsburg started in 1905, when a whaling station was established, but it closed in 1912. During that period, Norwegian whalers harpooned approximately 2000 larger species of whales (such as fin whales and blue whales) around Spitsbergen, and their populations have still not recovered. The first wireless telegraph station in Svalbard was built at Barentsburg, but moved to Longyearbyen in 1930. However, the main activity has been coalmining. The first mine was started in 1912 by a Norwegian company. In 1920 the mine was bought by the Nederlandsche Spitsbergen Compagnie, which founded a settlement named after the great, Arctic son of the Netherlands: Willem Barentsz. In 1932 Russia and Trust Arktikugol bought Barentsburg and the coalmine, much to the Norwegian Government's frustration. Barentsburg was evacuated in 1941 and largely destroyed, and the coal was purposefully set on fire. Serious fights between German, British, and Norwegian forces took place there. In 1946 mining was restarted, and the mines are still in operation today. Barentsburg has suffered from a range of minor difficulties (e.g. mine accidents, coal depots on fire) and accidents, such as a plane crash in 1996.

The number of buildings in Barentsburg is substantial, but the state of some of them is rather poor. Even the new yellow brick buildings seem partly dilapidated. Recent renovations have greatly improved some of the buildings and a few older houses have been removed. For example, the hotel was renovated recently and now appears very attractive, and the entire school building was beautifully decorated with colourful frescoes, reflecting not only major Russian sights but also Norwegian ones.

With the impressions of colourful Longyearbyen freshly in mind, arriving at Barentsburg provides an eyebrow raising contrast, from a modern, high-quality settlement, with distinctive buildings to a rundown-looking community dominated by black smoke from the power plant, old and redundant constructions, and dirt roads. Part of the contrast derives from the fact that there is an active mine in the centre of Barentsburg and that what was considered the normal view of industrial settlements everywhere 50 years ago has since changed in most parts of Europe. If one compares Pyramiden with Barentsburg, the former is a far more modern and well-organised settlement. Some years ago, when there were still many children present in Barentsburg, a couple of dirty cows, which produced milk for the children, could be seen wandering through the settlement in summer. Today, one can observe black and white cows in the Arctic with mixed emotions, 'grazing' where there is no grass. What on earth were the cows grazing on in the past? The now abandoned playgrounds and metal swings, a lonely, painted metal sunflower, and large fresco of summer-fresh birch trees touch the soul of some who have memories of home and the endless summers of the Ukraine. What words can we find for the mix of upwelling, discordant feelings?

After the dissolution of the Soviet Union and the fatal plane crash in 1996, the inhabitants of Barentsburg built a wooden chapel, a marvellous antithesis to the grimness of the engirdling socialistic expressions. The carefully carved wooden chapel is overlooked by the impressive, multi-storey brick building of the Russian Federation's General Consulate, which is encircled by a tall, solid metal fence. Two worlds stare at each other: one carved by the people in materials derived from living nature, and with round, playful shapes; the other built by a demonstrating state, in bricks, steel, and concrete, with a character that seems to display power and domination. As one approaches the consulate, it appears to watch over the entire settlement – an enclosed world engirdled by a tall fence. What is kept at a safe distance from the representative of the Russian State – its people, or possibly its visitors? The visitors from less top-down organised and less restricted societies ponder about these antagonisms and clashes of existence between power display and people. If we shape a building for and with the people and an integrated society, what architectural style should be selected to create an outlook that takes care of people's needs and invites them to experience cooperation and synergy?

What will be the future of Barentsburg? Obviously, the times of coalmining are passé, but what else can be done there? Will Barentsburg become a new 'Pyramiden'? This does not seem likely. Russia is strongly determined to be an Arctic nation, if not *the* Arctic nation, and probably will not abandon Svalbard. Will the future involve research, tourism, and commerce, as in the case of Longyearbyen? If so, Barentsburg's image will have to change; it will need a major facelift and a fundamental redefinition of its role.

When one strolls through the most often deserted streets of Barentsburg and passes living quarters, one is likely to be struck by the large number of noisy kittiwakes that nest on the window frames and on top of boxes that appear to be air-conditioning units. However, appearances are deceptive, because the metal boxes serve as simple refrigerators! Although there are few people in Barentsburg, there is lots of life. Human homes have turned into bird homelands. Thus, nature strikes back. ■

Smeerenburg

The settlement of Smeerenburg on Amsterdamøya in north-west Svalbard was first occupied by the Dutch in 1614, when ships from the Amsterdam Chamber of the Noordsche Compagnie established a temporary whaling station with tents and crude, temporary boilers. During the first intensive phase of the Spitsbergen whale hunting, Smeerenburg served as the centre of operations in the north. The name, in Dutch, literally means 'blubber town' or 'fat city'. In 1619, a 500-ton ship with timber and other materials was sent from The Netherlands to Spitsbergen. The tents and temporary ovens were respectively replaced with wooden structures, copper kettles, brick foundation and chimneys.

In its first year, only people from Amsterdam occupied Smeerenburg. In 1623, Basque ships employed by the Danes arrived at Smeerenburg and a time of turmoil began. By 1626 there were five big huts in Smeerenburg, and by 1633 all the chambers of the Noordsche Compagnie were represented at the settlement. In its heyday (1630s), Smeerenburg comprised 16–17 buildings, including a small fort at its centre. There were eight ovens situated in front of the buildings. During that time as many as 200 men were working ashore, flensing whales, boiling the blubber to extract oil, and coopering casks to contain the oil.

Following the destruction of a Dutch station in Jan Mayen by Danish-commissioned Basque ships in 1632, the Dutch sent seven men to overwinter at Smeerenburg in 1633–1634. All men survived, thus prompting overwintering by another group of seven sailors in 1634–1635. Unfortunately, all seven men perished and the practice of overwintering in Smeerenburg was abandoned. Smeerenburg's decline began in the 1640s. The settlement was abandoned around the year 1660, at the time when there was a transition to processing blubber into oil on return to the homeport and an expansion into pelagic whaling.

The size of Smeerenburg has been greatly exaggerated by many authors. William Scoresby (1820) claimed that 200–300 ships and 12,000–18,000 men visited Smeerenburg during the short summer season. Fridtjof Nansen (1920) made similar claims, stating that hundreds of ships anchored in the roads of Smeerenburg, where 10,000 people visited a proper town. Besides the workshops, there were shops, churches, fortifications, and even brothels. Such claims have no basis in reality. No more than 400 men would have visited Smeerenburg during the peak of whaling in the 1630s. There were no shops, churches, or brothels, although there was a single fort with two guns.

Remains and majestic views from the heyday of whaling on Smeerenburg (page 178). The hand-coloured lithography (Vue Prise dans la Baie de Smerenberg, Spitzberg) was made by Léon Jean-Baptiste Sabatier, after a lithography by B. Lauvergne (page 252). In the centre, there is a man with a polar bear fur, a woman, and a second man studying the view due north – respectively the La Recherche expedition leader Joseph Gaimard (left, page 215), the writer Léonie d´Aunet (centre, page 226), and probably either the artist François-Auguste Biard (page 250) or the writer Xavier Marmier. They are small figures in a grand, cold landscape. On the following pages: the view from Smeerenburg due north, studied by so many explorers who wished to reach the North Pole. ∎

Svalbard Life | 179

Virgohamna

A stone's throw away from Smeerenburg, we find Virgohamna (page 34). People that are interested in aviation should visit this site, which was as famous and media-encircled at the end of the 19th century as today's Cape Canaveral. For a period of 10 years Virgohamna became a prominent high-tech site for aviation, where the best possible technologies of the time were employed to reach the exalted, fabled, and ultimate goal of arctic exploration, the North Pole. However, there is also a much less mentioned prelude in Virgohamna: whale hunting.

The 'Harlinger kokerij' in what then was called Houker Bay (present-day Virgohamna), was established by the Dutch in 1636. Obviously, there were space issues in Smeerenburg. Already a few years later, whale hunting in the region had declined and in the 1640s the 'Harlinger kokerij' lost its significance, one of a number of typically short Arctic adventures, that in this case lasted only 10 years. The foundations of buildings and the ovens are clearly visible in Virgohamna.

Despite a few visits, historic silence sinks over the 'Harlinger kokerij'. History jumps about 240 year to the year 1888, when a rather eccentric Englishman, Arnold Pikes, decided to experience overwintering in Svalbard. Moreover, he did not decide upon anywhere 'convenient', but rather the most desolate, yet still accessible place imaginable. He arrived, together with a prefabricated house, in what today is officially called Virgohamna, but at that time was named Pikes Bay. He overwintered in the house, not alone, but in the company of six sailors that had erected the house. It must have been a rather

social overwintering compared to some others! Pikes becomes the first overwintering tourist in the archipelago. His house was later used by the explorers Andrée and Wellmann, and he showed up personally in the winter of 1896–1897 to experience what we today would call the countdown and lift off of Andrée's balloon.

This Arctic 'Cape Canaveral' started with the arrival of the ship the *Virgo*, which transported Andrée's expedition equipment. Thereafter, the place was known as Virgohamna (Virgo Harbour). Several buildings and a hangar were raised. A hydrogen plant was built, gas tubes installed, and chemicals for gas production stored. Thus, the technological wonder of a balloon was installed to reach a desired goal in the expanse of the Arctic ice cap. At the time, the place was characterised by what must have appeared to be the most outstanding high-tech equipment, and continuous streams of visitors, journalists, and tourists wanting to experience *the* event of the 1890s swarmed around the area. A most unexpected and sensational guest arrived from the north on August 1896, namely the battered and worn vessel the *Fram*, directly after it had broken free of the year-long embrace of the ice caused by the Transpolar Drift. They were able to inform that the *Fram* had not reached the North Pole, and upon hearing the news Andrée must have let out a sigh of relief: the North Pole was still untouched and there for the taking. However, the crew of the *Fram* had a more pressing question: had anybody heard from Nansen and Johansen?

This photograph records the remains of the times when Virgohamna was an essential place for Arctic explorers.

After Andrée and his men took off into their Armageddon, the American Walter Wellman (1858–1934) stepped onto the scene, introducing the next generation of aircraft: airships. Wellman was not an engineer, pilot, or scientist. He was a journalist that in 1894 had visited Pike's house during a voyage on the Tromsø-based ship the *Ragnvald Jarl*. The trip had inspired him to attempt to reach the North Pole by air (after he had failed to reach it by aluminium ships and dog sledges). In 1906 he arrived in Virgohamna with a brand new airship and formidable financial resources for the settlement. The biggest airship in the world was built in Paris, and for this Wellman had assembled a hangar (58 m long and 26 m high), a workshop, and a hydrogen plant. Nothing really had gone to plan in 1906. When he came back in 1907 to terminate the unfinished hangar it was blown down on the American national day! However, by September in the same year, far too late in the season, everything was ready and the airship made its first trip in the Arctic. It landed on a glacier a few hours later.

Surprisingly, the damaged airship was saved. The final attempt to reach the North Pole was made in 1909. While heading due north, the airship suddenly lost a 500 kg heavy drag wire and rose from its planned flying altitude of 80 m to 1800 m. A valve was opened and the airship sank to the ocean surface. Those on-board were lucky! The Norwegian research vessel the *Farm*, under the command of Gunnar Isachsen, was directly below them! The aeronauts were rescued. They moved the airship back to Virgohamna and with 200 tonnes of hydrogen above them Wellman ignited one of his famous cigars, celebrating his survival. The inevitable happened, and the new accident resulted in the airship losing too much weight and climbed higher and higher until it exploded and rained down in pieces over Virgohamna and the *Farm*. It is said that Wellman had opened a bottle of champagne and played a card game of patience after the event. Nansen had no mercy for Wellman and called him a 'humbug'.

After the end of the 'Wellman' period, life became quiet in Virgohamna, but in 1928 things picked up again. In common with Ny-Ålesund, Virgohamna had become a base for rescue teams attempting to rescue the Italian explorer and airship pilot Umberto Nobile and his crew from the ice. Again, there was full media coverage, but for the last time – so far. Since then, silence has prevailed, except for one or other tourist group that pays a visit. No one is allowed to go ashore without specific permission obtained from the Sysselmann. One can study how a prime site of human endeavours, the Arctic Cape Canaveral, was turned by the climate (in addition to theft and pillage) into an archaeological site.

This account of Virgohamna is more of a sketch based upon history and stories than a description of a place. The combination of stories, wreckage, and a wealth of metal, ceramics, and planks can conjure up an imagined site with large hangars, gas plants, engineering workshops, and stores of chemicals. Today, Virgohamna is a sea of trashed items, refuse, and carcasses, but in our imagination it can be a place where the most modern technology was applied, at the edge of the possible, to fulfil the dreams of so many: to reach the North Pole.

Houses and engines in Virgohamna in the late 19th century (left) and a similar view today (page 185). Wellmann's airship is tested over water (right). ■

Ny-Ålesund

The northernmost year-round settlement in the Arctic is Ny-Ålesund (page 34), apart from a few permanently staffed military and meteorological stations in Canada and on Greenland. Ny-Ålesund is situated on the southern shores of Kongsfjorden and enjoys great vistas of the mighty glacier Kongsbreen and the triple mountain peaks named Tre Kroner. Behind the settlement, that has approximately 35 inhabitants in winter and approximately 200 in summer, the famous mountain Zeppelinfjellet (Zeppelin Mountain) is situated, housing one of the most significant observation sites for global climate and air pollution. Ny-Ålesund even has a small airport for connecting flights to Longyearbyen.

In Ny-Ålesund, as in other settlements in Svalbard, coalmining was the main reason for the development of the settlement. After a cautious start, the still existing Kings Bay Kull Co A/S (KBKC) was founded in 1916. The coal was in a narrow seam, with significant fault lines, situated several hundred metres below sea level. An entire settlement, infrastructure, and mine had to be built up from scratch. Methane release resulted in serious accidents. Several hundred people were working there, but in 1929 mining ceased and the Norwegian state bought all shares. In the 1930s unsuccessful attempts were made to turn Ny-Ålesund into a fish plant.

Despite the difficulties, Ny-Ålesund became world famous due to the dramatic attempts to reach the North Pole. Roald Amundsen, who had not been able to reach the North Pole by ship (the Maud expedition through the Bering Strait), made his second attempt from Ny-Ålesund with two Dornier planes in 1925. The entire crew survived under dramatic circumstances, but the magic goal was not achieved. Amundsen, a 'real man', i.e. one that never gave up, returned to Ny-Ålesund in 1926 with the airship *Norge*, under the command of the Italian Umberto Nobile (1885–1978). The airship's mooring mast is still in place. At this third attempt, Amundsen, Nobile, and their crew flew over the pole to Teller, Alaska, without encountering any difficulties. Although Nobile was essential in this victory, the all-overshadowing Amundsen went off with most of the prestige. Should the fame for this climax of attempts to reach the North Pole have gone to the young, recently independent nation Norway or the ancient and, at the time, ultra-nationalistic, fascistic Italy? Nobile returned to Svalbard to reclaim Italian honour and pride. In 1928

Nobile returned to Ny-Ålesund with the airship *Italia*, which, after reaching the pole, crash-landed on the ice, resulting in the biggest ever rescue operation in the Arctic (20 aircraft, 14 ships, and 17 casualties), with the entire world following the news, such was the dramatic end of the heroic age to reach the North Pole. Both tiny Ny-Ålesund, and Virgohamna, were right at the heart of these events. The house in which Amundsen lived has recently been renovated and exploration paintings by an unknown artist were recovered.

After evacuation early in World War II, mining started again in 1945. By 1960, the population had reached approximately 150 men, 50 women, and 40 children. In 1953, 10% of the population lost their lives in two mine explosions. A further major explosion in 1964 resulted in the final closedown of the mine. This tragedy generated a major crisis that changed the political scene of Norway for good.

The view of the Ny-Ålesund airport from the west. Right: A view of Zeppelin Mountain.

Whalers were familiar with Kongsfjorden already in the 17th century, after Jonas Poole had found pieces of coal there in 1610. The Pomors also used the fjord as base for hunting. The famous British whaling captain William Scoresby junior made the first description of the region in 1818. Already in 1910, topographic and meteorological research was carried out by Ferdinand Graf Zeppelin in order to investigate whether Ny-Ålesund could be used for regular airships. Norway operated a geophysical research station from 1920 to 1924. After World War II, trans-Arctic flights demanded the building of airfields, and the possibility of building one in Ny-Ålesund was evaluated. In the middle of the Cold War, protests by the Soviet Union, fearing that the airfield could be used for military action, became too strong and the plans were abandoned. However, the excellent position of Svalbard and Ny-Ålesund for satellite-based telemetric stations became apparent and Norway built its first station in 1964. Following this, a range of state-of-the-art research instalments and institutions were established, thus paving the road for all further development of Ny-Ålesund up to the present day.

Here, I mention just a few of the institutions at Ny-Ålesund. The Norwegian Polar Institute has been present year-round since 1968. There is a geodetic station run by the Norwegian Mapping Authority. Scientific rockets are launched by SvalRak. Germany built the Koldewey Station, which is run in concert with France and studies, inter alia the ozone layer. Also present is the Japanese National Institute of Polar Research, the English National Environment Research Council, the Italian Consiglio Nazionale delle Richerche, the South Korean DASA Station, the Chinese Yellow-River Station, and the Indian Himadri Station. One of the latest institutions to be established in Ny-Ålesund is the Kings Bay Marine Laboratory, a unique experimental research location with running seawater for all-year experiments on Arctic marine organisms.

The history of Ny-Ålesund is thus manifold, with a rich mosaic of dissimilar roles. It is a place of dramatic events and rapid shifts, but continuous attention.

The cable car that takes scientists to the time-series climate and pollution instruments on Zeppelin Mountain (upper left and centre). Ny-Ålesund, with its characteristic skyline, painted by Gunnar Wefring in 1936 (lower left). The historic photograph of the airship *Norge* lifting off from Ny-Ålesund in 1925, en route for the North Pole, with Amundsen and Nobile on-board. ∎

Ny-London

In the early days of exploration, the exploitation of minerals such as coal, marble, and gypsum were common in Svalbard, and a large number of international companies were operating there. One of these was the Northern Exploration Company Ltd. (NEC), which received funding from London in 1910. The company provided financial support for continuation of earlier endeavours to mine marble. NEC mined minerals in several places in Svalbard, but never accumulated any profits. In 1911 they developed a new mining site on the northern shores of Kongsfjorden, called Ny-London (page 34). The settlement had houses (now located in Ny-Ålesund), workshops, a quay, and mobile equipment for generating steam power. Marketing efforts for marble of various qualities were substantial. However, only a few tonnes of marble were produced and were completely unsuitable for building purposes. Despite this, NEC continued the activity and more capital was raised and invested! No company ever invested so much money in Svalbard in that period as NEC.

By 1920 NEC had reported to the British foreign department occupational claims of 10,000 square kilometres, i.e. one-sixth of the entire area of Svalbard. However, the level of activity was low and attempts were made to sell NEC to the Norwegian state. By 1929, NEC was close to bankruptcy and in 1932 the land owned by NEC was sold to the Norwegian state, which subsequently dissolved the company in 1934.

The story of NEC and Ny-London is one of many that shows how difficult it is to run a successful business in Svalbard. The remains of countless attempts to exploit natural resources can be found along the Svalbard coastline. From the time dominated by the Dutch, there are the remains of whale oil production sites, settlements, and graves. Similarly, from the 'Pomor period', succeeded by numerous endeavours such as the NEC the strategy has always been the same: (try to) go in, (try to) get rich, and (certainly) get out! Few have become rich in Svalbard, but many have sustained their way of life. The remains of past exploiters are left behind, as symbols of false aspirations and unsuccessful endeavours. Today, we would demand clean-up actions, but the remnants of the past have now become cultural monuments that represent bygone activities.

The remains of machinery and housing are visible, as well as disintegrated marble. ∎

Svalbard Life | 191

Svea

Svea is a mining settlement at the head of the enclosed Van Mijenfjorden (page 34). It is the third largest settlement in the territory (after Longyearbyen and Barentsburg). Currently, around 300 workers live in Longyearbyen and commute to Svea to work on a daily or weekly basis. There are no permanent inhabitants living in Svea. The settlement is operated by SNSK and served by the small Svea airport.

Swedes first established the town in 1917. It was destroyed in 1944, but was quickly re-established after WWII. Mining activities ceased in 1949, and were not re-established until 1970. The town almost vanished, due to more productive and accessible mines in Longyearbyen. Mining was suspended for a short period in 1987, and also noteworthy is a mine fire in 2005, which lasted uninterrupted for more than five weeks and caused major damage.

Today, Svea has the most productive coalmine of Svalbard, the Svea Nord longwall mine. The mine, which was opened in 2001, currently produces up to 4 million metric tons of coal annually, making it one of the largest underground coalmines in Europe. It is a modern mine, with the type of working conditions and amenities that miners in earlier times could only have dreamed about, including having a home, one's partner and children present, and the many conveniences of nearby Longyearbyen. ■

Svalbard Life | 193

The fjord named Hornsund is 30 kilometres in length and cuts through different geological formations, dating from the Precambrian in the west to the upper Mesozoic in the east. It is one of the most spectacular fjords in Svalbard, with majestic mountain peaks and magnificent wide glaciers (page 34). The English explorer Jonas Poole visited Hornsund in 1610, and named the fjord after fellow explorers had brought back a reindeer antler. In 1613 the first whaling ships used Hornsund. The English held almost a monopoly on whaling activities in the bay, until they abandoned it in the late 1650s, but also Dutch and Danish whalers fought for dominance there.

The Polish Arctic research station, the Polish Polar Station, is located at Isbjørnhamna in Hornsund. A Polish Academy of Sciences expedition erected the sta-

Hornsund

tion in 1957 within the framework of the International Geophysical Year. The station started operations under the leadership of Stanislaw Siedlecki, geologist, explorer, and climber, and a veteran of Polish Arctic expeditions in the 1930s (including the first traverse of Spitsbergen). The station was modernised in 1978, in order to resume year-round activity.

The contribution of Polish scientists to the investigation of Svalbard has been considerable, particularly with regard to the local glaciology, oceanography, marine biodiversity, seabird ecology, and marine ecology of western Spitsbergen. The work started out with the simplest means, but with steady dedication. In the difficult political years in Poland, before the dissolution of the Soviet Union, many of the scientists belonged to a part of the population that one could call dissidents. Demonstratively, a cross was erected at the entrance to Hornsund during those days, to signal to passing cruise vessels that Poland was more diverse than the country's official, communistic, and materialistic self-representation. With passion and dedication, the station was run often under difficult supply circumstances, but always with the declared support of the Governor of Svalbard. To enter the station as a visitor is a pleasure. The premises are presented openly and with pride, always accompanied by coffee and assorted cakes, or even a three-course meal. The old tradition of welcoming guests is taken care of with fervour!

Arctic research would be different if more European countries were to dedicate themselves as genuinely as the Polish Academy of Sciences. Most European polar research is not carried out in the nearby Arctic, subjected to the most significant greatest warming, but in far away Antarctica, where the effects of climate change are – hitherto – small. There is a range of historical and political explanations for this tradition, which has resulted in the Arctic Ocean being among the least-investigated regions of the world. However, is it a wise strategy to give lower priority to research in the Arctic? What can be our arguments for not emphasising global change research where it is breathtakingly prominent? Reason and reality do not always go hand in hand. If there were more stations or research ships on or around Svalbard, of the type exemplified by Poland, much would be different. ■

Hopen

The island of Hopen is situated at 76.30°N 25.01°E at the eastern edge of the Svalbard archipelago and the Barents Sea (page 34). Hopen is a narrow and long, rocky island of 46 square kilometres, with mountains 150–370 metres high. There are neither rivers nor lakes on the island, and freshwater derives from melted snow or rain. For approximately half of each year, Hopen is engulfed by drifting sea ice and is only accessible by helicopter. Hopen is an important landing and refuelling station for helicopters, particularly during rescue operations. The Norwegian Coast Guard transports supplies by ship during summer. Hopen has no harbour and all supplies have to be transported ashore by small ships. In the winter, helicopters bring minor supplies and post. The Norwegian Meteorological Institute operates a manned weather station on the island with a staff of four persons. The station is well equipped, spacious, and cosy. It also has six huskies, which keep watch for polar bears that frequently visit the station in search of food or out of curiosity. When spring arrives, an amazing number of birds come to brood in the stiff cliffs. Winter silence is broken and the sound levels are almost deafening at times.

Hopen was discovered in 1613, probably by Thomas Marmaduke of Hull, who named it after his former command, the *Hopewell*. The weather station, which is located on the south-eastern part of the island, was founded in 1946, and has been in continuous operation since then. During World War II, the Luftwaffe placed a meteorological team there. Also, a Soviet ship ran aground, with many casualties.

During spring and early summer, many polar bears leave their hibernating sites at Kongsøya and move with the pack ice southwards towards Hopen. This results in hundreds of annual polar bear sightings at the station. Like in a western saloon, a rack of loaded rifles, ready to be grasped, is placed at the entrance of the station during the period when the island is enclosed by ice. A polar bear may stand right outside! Some of the meteorological instruments are typically placed in characteristic white boxes outside and have to be checked regularly. One has to be daring to put one's head into such a box when potentially there may be a polar bear in the vicinity, especially when it is pitch-dark! The only support comes from the barking huskies on watch duty. It is not a job for town freaks or couch-radical nature lovers, whose love for wild nature rests at a safe distance from true wilderness.

For the welfare of the crew, three cabins are available on the island. Welfare? Cabin visits? One of the peculiarities in northern countries is that one has to experience nature as purely as possible, and in loneliness. Thus, making trips to huts is normal, even for four persons that are isolated in one building for six months. Whereas the desire to be totally alone and to draw a line with one's skis through virgin snow may be considered a symptom of psychological instability and sickly introversion elsewhere, this is nothing less than normal in the High North. Hence, the four members of staff based at Hopen Station chose to take their cabin trips so that, finally, they can be alone!

The station has its own 'Aqualand' not far from the station. It comprises of a spacious sauna and an outdoor wooden tub that can be heated with wood (logging along Siberian rivers results in plenty of driftwood all over Svalbard). This mini water park is often used when, occasionally, the station receives visitors. On one occasion, two visitors had decided to enjoy life in the outdoor

wooden tub. The evening skies were clear and the stars were twinkling. While relaxing, they had gazed through the clouds of vapour rising from the water surface to view amazing scenery. During that tranquil time they suddenly became insecure as to whether they were having an apparition or were facing reality. Both glared into the dark surroundings. They had not seen an Arctic mirage, but a polar bear passing by – between the tub and the station. The rifles were safely stored inside the 'aqualand' building. They were faced with a dilemma about what to do – whether to dive down and enjoy the warm water, hoping for the best, and risk the possibility of becoming two dumplings in a thin soup. Often in such situations, the best option is to wait. In the event, nothing happened and the bear continued on its way. Clearly, even a peaceful bath in an outdoor wooden tub can become a thrilling experience under the stars and Northern Lights of an Arctic sky. Life in Svalbard may provide a plethora of memorable experiences that seemingly add new depths to the everyday life that most people experience. No wonder that, for some, life in the wilderness develops into an infectious disease!

Waters with melting ice off the coast of Hopen (page 189) and the welcoming station interior. On pages 200–201, the head of the station in 2010; Kåre Holter Solhjell, accompanied by one of the essential huskies. ■

Bjørnøya

The island of Bjørnøya is the southernmost island of the Svalbard archipelago (page 34). It has an area of 178 square kilometres and the maximal altitude is approximately 540 metres. The island is located in the south-western part of the Barents Sea, approximately halfway between Spitsbergen and the North Cape. The Dutch explorers Willem Barentsz and Jacob van Heemskerk discovered Bjørnøya in 1596. It was named after a polar bear that was seen swimming nearby. Steven Bennet conducted further explorations in 1603 and 1604 and noted the rich population of walruses. Starting in the early 17th century, the island was used mainly as a base for the hunting of walrus and other seal species, but these rich resources were soon depleted. Seabird eggs were harvested from the large bird colonies until 1971. Despite its name (Bear Island), the island is not a permanent residence of polar bears, although many arrive with the expanding pack ice.

Because Bjørnøya lies on a boundary between cold Arctic and warmer Atlantic waters, water temperatures around the island are quite variable, sometimes reaching 10°C in summer. During the winter, fast ice develops along the coast. The Barents Sea carries pack ice to Bjørnøya every winter, sometimes as early as October, but significant amounts of ice are not common before February.

The island was considered no-man's-land until the Svalbard Treaty placed it under Norwegian sovereignty. Despite its remote location and barren nature, the island has seen commercial activities in past centuries, such as coalmining, fishing, and whaling. However, no settlements have lasted more than a few years. From 1916 to 1925, coal was mined at a small settlement, but mining was given up as unprofitable. Bjørnøya is now uninhabited except for the personnel working at the island's meteorological station, Bjørnøy Radio. The station is used as a base for meteorological observations and for providing logistics and telecommunication services, including radio watch duties. Weather forecasts are transmitted from the station twice daily. The station also maintains a landing platform for helicopters. A small number of cruise ships visits the island, but otherwise tourism is almost non-existent. Together with the adjacent waters, it was declared a nature reserve in 2002.

The large photograph shows Bjørnøy Radio's harbour, with the research vessel the *Helmer Hanssen* on the horizon. On page 203, visitors in a rubber boat are welcomed to the island. The tall rock is part of the mountain named Perleporten, at the southern tip of Bjørnøya. Below, there is a building close to the tarmac used by helicopters and a view of the island from the sea. ■

VII. PEOPLE: THEN AND NOW

PEOPLE: THEN
EXPLORERS
SCIENTISTS
HUNTERS
WOMEN
MINERS

PEOPLE: NOW
ODD OLSEN INGERØ
LENA ROMANENKO
ROGER JACOBSEN
HANS ROAR HANSEN
FIONA DANKS
CARLOS DUARTE
TOVE GABRIELSEN
OLE MAGNUS RAPP
JAN MARTIN BERG
JOHNNA HOLDING

Explorers:
Prince Albert I of Monaco

Among the explorers of Svalbard, Prince Albert I of Monaco (1848-1922) played a particularly important role. The prince organised and financed four expeditions to Svalbard and contributed significantly to the development of Norwegian research on Spitsbergen, which led to the foundation of what would become the Norwegian Polar Institute. At 18 years old, Prince Albert I joined the Spanish Navy and eventually became a captain. Later, he joined the French Navy. From 1870 and onwards he participated in several scientific expeditions and became an oceanographer. Through most of his life Prince Albert I commissioned and headed oceanographic cruises to the Atlantic Ocean and the Mediterranean Sea. For this purpose, he arranged to have modern research ships built, which were some of the first specifically designed for oceanographic studies. His expeditions to Svalbard are of particular significance for the history of the High North, as they were very well organised and always international. His first two expeditions took place in 1898 and 1899, mainly dedicated to oceanography. His main thrust in Svalbard research was through his expeditions in 1906 and 1907. In addition to the Prince's personal working consortium, two other groups of scientists were present: a Scottish group headed by the marine biologist William Speirs Bruce (1867-1921), and a Norwegian team working on mapping and topography and headed by Gunnar Isachsen (1868-1939). Isachsen concentrated his explorations in the north-western sections of Spitsbergen, today called Albert I Land. In 1907, the famous geologist Adolf Hoel and botanist Hanna Resvoll-Holmsen joined Prince Albert I's expeditions. The topographical studies, the efforts to map glaciers, and some of the first botanical mapping became essential for Norwegian science in the archipelago. The Prince embarked on the Svalbard cruises with his biggest research ship, the *Princesse Alice II*, which was partly supported by smaller support ships.

The scientific research was funded by the economic wealth that Prince Albert I had accumulated in his small, but increasingly more prosperous principality of Monaco. As a dedicated scientist and a person that wished to reach out to people, all expeditions were accompanied by numerous artists that painted or sketched landscapes, and made coloured drawings of marine animals. Colour is an important feature of marine animals that is easily lost after death or if they are collected and stored. The results of few marine expeditions were so well presented with photography, graphics, and art than the expeditions of Prince Albert I. Among the artists were Louis Tinayre and Witold Lovatelli-Colombo, which are present in more detail in chapter VIII. The Prince made sure that the results of the expeditions were thoroughly published in five volumes of *Résultats des campagnes scientifiques accomplies sur son Yacht par Albert Ier*. Prince Albert I also recognised the need to reach out not only to scientists but also to particularly interested members of the general public. The Musée océanographique de Monaco, which was built under the orders of the Prince, is a breathtaking building on a steep headland cliff that, in a mixture of modern and old styles, houses one of the world's best oceanographic exhibitions. To date, the L'Institut océanographique de Paris (Institute of Oceanography), founded by Albert I in 1907, has been a prime site for meetings, conferences, and teaching.

Both Hoel and Isachsen thanked Prince Albert I for his significant contribution to Norwegian research especially in 1906 and 1907, without which Norwegian research in the Svalbard region would have had a far more difficult start. The Prince's unselfish, international, and science-focused approach created the base for future Norwegian research in the region.

The photograph on page 209 shows Prince Albert I in navy uniform during an early phase of his reign. He has the appearance of a man of action, as though he were about to ask 'Don't you want to join me on my next expedition? Don't you wish to learn more about hidden life in the depths of the ocean?' The small photograph, taken onboard the research vessel the *Princesse Alice II* off the coast of Svalbard, shows Louis Tinayre (second from left), Gunnar Isachsen (centre), and Prince Albert I (to the right). ■

208 | Svalbard Life

450. Spitsbergen: Andrées Station paa Danskøen. G. Hagen, Hammerfest.

Explorers: Salomon Andrée

Salomon August Andrée's Arctic balloon expedition in 1897 was an ill-fated effort to reach the North Pole, in which all three members of the expedition perished. Andrée (1854–1897) was the first ever Swedish balloonist. He proposed a voyage by hydrogen balloon from Svalbard to either Russia or Canada, which was to pass, with luck, straight over the North Pole on the way. His expedition was one of the hottest media events in 1897 and tourist ships appeared at Virgohamna, where the expedition started. Journalists were present and artists and professional photographers produced photographs, graphics and paintings of the event. The scheme was received with patriotic enthusiasm in Sweden, a northern nation that had fallen behind with regard to Arctic expeditions and the race for the North Pole, a desolate point that even today receives great attention. 'Who owns the North Pole' is a question that is vividly discussed and ownership is sought after by the Arctic coastal nations Denmark/Greenland, Canada, the USA, and Russia. But not Norway, being the only Arctic costal nation that has its sea borders, including the Arctic ones, defined and internationally accepted.

Andrée neglected many signs of the dangers associated with his planned venture. Being able to steer the balloon to some extent was essential for a safe journey. There was plenty of evidence that the drag-rope steering technique he had invented was ineffective, yet he staked the fate of the expedition on drag ropes. Worse, the polar balloon Örnen (Eagle) was delivered directly to Svalbard from its manufacturer in Paris without having been tested. When measurements showed that it was leaking more hydrogen than expected, Andrée refused to acknowledge the alarming implications. Most modern students of Andrée's expedition see his optimism, faith in the power of technology, and disregard for the forces of nature as the main factors in the series of events that led to the deaths of the entire team.

After Andrée shouted 'cut off everything' in July 1897 and the balloon disappeared with southerly winds over Smeerenburg into the Arctic Ocean, he was literally cut off from the world and flew into his self-selected Armageddon. Soon after lift-off the balloon unavoidably lost hydrogen quickly and crashed on the pack ice only two days later. The explorers were unhurt but faced an exhausting trek back south across drifting ice. Inadequately clothed, equipped, and prepared, and shocked by the difficulty of the terrain, they did not make it to safety. As the Arctic winter closed in on them in October, the group ended up exhausted on the deserted island of Kvitøya in north-eastern Svalbard and died there.

For 33 years, the fate of the Andrée's expedition remained one of the many unsolved riddles of Arctic exploration. By chance, the Norwegian sealing ship Brattvåg discovered the expedition's last camp in 1930. It created a second media sensation, this time in Sweden, where the dead men were mourned and idolised. Some of the equipment (including roubles and US and Canadian dollars!), a logbook, and a camera were discovered. The fate of the explorers could thus be reconstructed. Andrée's motives have since been re-evaluated, as is the case for all polar heroes. The polar regions have long been considered the proving ground of masculinity and patriotism. Masculinity, heterophily, and polar heroes are still clearly combined entities in our conceptions. When the Danish author Klaus Rifberg suggested a homo-erotic relationship between the Arctic heroes Nansen and Johansen, a hefty debate took place, even at times when the entire range of sexual tendencies are a natural segment of public domain.

Today, we see Andrée more in the light of a weak and potentially cynical character, exposed to the mercy of his sponsors, Swedish patriotism, the media, and his time. Modern writers' verdicts on Andrée's virtual sacrifice of the lives of his two younger companions vary in harshness, depending on whether he is seen as the manipulator or the victim of Swedish nationalist fervour around the turn of the 20th century. The portrait of Andrée appears to encompass the characteristics of self-confidence, determination, the possession success, arrogance, and brilliance.

On page 210, the balloon hangar is shown during the preparations for lift-off. See also the painting by Hans Beat Wieland made in 1896 (page 257). ∎

Explorers: Roald Amundsen

Roald Engelbregt Gravning Amundsen (1872–1928) was one of the greatest internationally renowned Norwegian explorers of polar regions. Along with Fridtjof Nansen, Robert Falcon Scott, and Ernest Shackleton, he was a key explorer during the heroic age of polar expeditions. Here, we focus upon Amundsen's Arctic endeavours from their base in Svalbard.

In 1903, Amundsen successfully led the first expedition through Canada's Northwest Passage between the Atlantic Ocean and Pacific Ocean. So many had tried before, but with six others and the small, 47-ton steel seal-hunting vessel the *Gjøa*, from Tromsø, Amundsen succeeded. Probably, he was successful because he had learned from his fellow countrymen in Norway, which in those days was badly off, that small is beautiful, that one cannot apply force when it comes to nature, and that adopting the lifestyle of the Inuit is a precondition for survival in the High Arctic. The latter tradition was introduced in Norway through the unplanned overwintering of Nansen in present-day Nuuk, after the first crossing of the Greenland ice cap. While traversing the Northwest Passage, Amundsen learned additional lessons from the local Netsilik Inuit about Arctic survival skills, which would later prove useful. For example, he learned to use sledge dogs and to wear animal skins instead of heavy, woollen parkas. Those that were arrogant enough not to adopt a quasi Inuit lifestyle in polar regions either did not survive or only survived with difficulties. After several winters trapped in the ice, Amundsen was able to navigate into the Beaufort Sea and thereafter the Bering Strait.

In 1925 Amundsen and a team of five men took two Dornier flying boats to a latitude of 87° 44'N – the northernmost latitude reached at that time. The aircraft landed a few miles apart without radio contact, yet the crews managed to reunite. One plane was damaged and Amundsen and his crew worked for over three weeks to clear an airstrip ready for take-off with the remaining plane from the ice. They shovelled no less than 600 tons of ice on 400 grams of daily food rations. In the end, six members of crew were packed into the remaining plane. They barely managed to become airborne over the cracking ice and then had to make an emergency landing in northern Svalbard, where they were eventually rescued by a passing vessel. When it was thought that the crew members had been lost forever, they returned to Ny-Ålesund, triumphant. In 1926, Amundsen and 15 other men made the first crossing of the North Pole in the airship *Norge*. The three previous claims to have arrived at the North Pole are all disputed as being either of dubious accuracy or of outright fraud. Amundsen disappeared on 18 June 1928 while flying on a rescue mission to look for missing members of the Italian explorer Nobile's crew.

Amundsen was undoubtedly one of the greatest polar explorers. He showed that one could attain the sheer impossible with simple means, but with the right attitude, the right preparation, and a cool, humble mind when it comes to nature. With dry and uncompromising harshness, he stated: 'Victory is waiting for the person that has everything well prepared – we call it luck. Defeat is an unconditional consequence for the person that has not taken the necessary steps in due time – we call it misfortune.' He had a difficult, rather demanding character and one can question whether he had true close friends or was capable of developing such friendships. Amundsen's character seemed only apt to reach his goals or perish in the attempt. He was not the dream son-in-law to have around during family gatherings. After his death, he became one of Norway's biggest heroes. Now that the Arctic is no longer the proving ground of masculinity and patriotism, he can be regarded as less heroic. His polar explorations were highly significant for the political self-esteem of Norway, which became independent as late as in 1905, while he was sailing through the Northwest Passage. He never contributed significantly to science, in contrast to Nansen, but was a great explorer and a master of logistics. He was (and remains) a controversial figure, but in his time he was important for Norwegian identity building.

The portrait of Amundsen (right), painted in 1935 (7 years after his death) is on display at the Norwegian Polar Institute. It was done by the Norwegian artist Astri Welhaven Heiberg (1881–1967), who is best known for her images of naked women in landscapes. She was neither met with support nor understanding and over time she gradually focused more and more on portraits. In the portrait, the great Amundsen, meticulously dressed, appears to be looking at something outside a window, or perhaps he is shown in a moment of introspection while he is fastening his eyes onto a landscape? Alternatively, as one of the most famous celebrities of his time, he may have been on his way to a dinner party? His ascetic, quasi-relaxed posture, eagle-nosed profile, determined expression, and killing eyes make him the unique person that he was: an independent, ambitious, firm, determined and authoritarian person. Possibly, he was unable to communicate in an empathic manner with people in his vicinity. He was the only one of his kind, but a lonely wolf. At the time, Norwegians wanted to place Nansen on a pedestal as an Arctic hero, but Nansen had other and very different roles: he was fond of people and too much of a humble humanist to become a stylite or pillar-saint (a type of Christian ascetic who in the early days of the Byzantine Empire stood on pillars preaching, fasting, and praying). The situation with Amundsen was different. He could only exist as a stylite, uplifted from the ordinary, closer to a heaven of immortal heroes, but unapproachable and holed-up in the armour of his soul. ∎

Scientists: Joseph Gaimard

At the start of our small selection of scientists who have worked in Svalbard, we honour the achievement of Joseph Paul Gaimard (1796–1858). He was the leader of all legs of the La Recherche expedition (1838–1840), the French-financed science expedition that explored the Færø Islands, Iceland, Norway, Spitsbergen, and the south-eastern Barents Sea. Gaimard was a French military surgeon and scientist. He circumnavigated the globe twice, and in 1835 and 1836 he travelled to Iceland with the corvette La Recherche, after which the expedition was named. Gaimard convinced King Louis-Phillipe, who wished to strengthen the position of France in science, to finance the rather extensive expedition to explore Europe's northern regions. Gaimard asked his friend François-Auguste Biard to join the expedition as one of its main artists, and in turn, Biard convinced Gaimard to accept his 18-year-old fiancé, Leonie d´Aunet, on board.

Research does not only need bright heads, but also good organisers. The idea of the lonely scientist that, by his mental capacity and in the remoteness and solitude of his or her personal ivory tower, changes the world is a romantic one that rarely reflects reality. Most research is done by groups and teams that first and foremost need organisers to realise bright ideas. Such organisers often are not the most brilliant researchers themselves, but can see the progress of science in context. Gaimard was one such great organiser and all credit is due to him for changing the view of the European Arctic. On well-placed feet on the ground of Arctic realities, he perceived modern research at the horizon of time: integrated, multidisciplinary, international, and addressing the big research questions.

Here, Joseph Paul Gaimard is shown depicted by François-Auguste Biard (see page 250), as part of the lithograph entitled *Vue de l´Ile de l´Ours ou Beeren-Eiland* (View of Bear Island or Beeren-Eiland (the original Dutch term for Bjørnøya)). The brave Gaimard is depicted with the picturesque skin of a polar bear over his shoulders, examining the sea in a rather melodramatic manner. What could he be looking for? In the background, the specific rock formation named Perleporten, on Bjørnøya, is covered with seabirds, while Gaimard's crew struggle to get their boat ashore. Gaimard's continued achievements stand behind the success of the La Recherche expeditions. ∎

Scientists: Fridtjof Nansen

Fridtjof Wedel-Jarlsberg Nansen (1861–1930) was a Norwegian explorer, scientist, diplomat, humanitarian, and Nobel Peace Prize laureate – four full lives were lived by one person. What can one write briefly that is strikingly new after all these years of learned books and essays regarding this 'grand homme'? Volumes have been filled on the subject of Nansen, who was highly professional in representing traditionally separated professional pillars, not only from the small nation of Norway's point of view, but also internationally. Here, we focus upon his role in Arctic science, with a brief excursion into his efforts as a humanitarian.

Nansen studied zoology at the Royal Frederick University in Oslo and later worked as a curator of the Bergen Museum, which in those days was a world-class institution in terms of zoology. His research on the central nervous system of lower marine animals earned him a doctorate after hefty debates regarding what were then considered controversial findings. However, today, we know that his research helped to establish modern theories of neurology. He was a champion skier and led the team that made the first crossing of the Greenland ice cap in 1888. At the age of 18 years, he broke a world skating record, and in the following year he won the national cross-country skiing championship, a feat he repeated on 11 subsequent occasions. Through his achievements, he introduced an era in Norway that led to the country's identification with winter sports.

Nansen's decisive endeavour in the borderland between exploration, science, and amazing courage was the North Pole expedition with the ship the Fram (1893–1896), which was especially built for the voyage. In the 19th century the spectacular and breathtaking expedition was the equivalent of the first landing on the moon. Not only did the expedition reach a record latitude of 86° 14' N, but also Nansen had the courage to leave the only secure point in the icy oblivion of the Transpolar Drift, namely the Fram. Together with Hjalmar Johansen, he attempted to reach the North Pole on skies, by dog sledge, and by kayak. The venture was not successful, and the two men returned and then survived the necessary overwintering in a stone cave on Franz Josef Land. Due to a diet based on seal and polar bear meat (including drinking blood and eating intestines – skills learned from Inuits), they did not develop scurvy, although they did put on much weight during the overwintering). What words can one choose to characterise their survivor skills at the utmost edge of the sheer impossible?

Nansen retired from exploration after his return to Norway. His endeavours as an explorer have since been overshadowed by his epoch-making work in oceanography. He became a professor at the University of Oslo, which gave him a base from which he could tackle the major task of editing the reports of the scientific results of the Fram expedition. The results were eventually published in six volumes. According to Robert Rudmore-Brown, at the time, the volumes were to Arctic oceanography the equivalent of what the results of the Challenger Expedition (which took Charles Darwin around the world) were to biology. In 1909, Fridtjof Nansen and his co-author Bjørn Helland Hansen published the milestone book The Norwegian Sea: Its Physical Oceanography. In the course of his research, Nansen embarked on many scientific cruises, mainly in the North Atlantic, and contributed to the development of modern oceanographic equipment. Nansen's retirement from polar exploration was the decisive step for the release of the Fram to other researchers. His retirement was significant for Roald Amundsen, who never became a scientist but undertook further North Pole expeditions.

As one of his country's leading citizens, in 1905 Nansen spoke out for the end of Norway's union with Sweden. He was instrumental in persuading Prince Charles of Denmark to accept the throne of the newly independent Norway. Between 1906 and 1908 he served as the Norwegian representative in London, where he helped negotiate the Integrity Treaty that guaranteed Norway's independent status. Nansen's contribution to his country's independence was extraordinary, and few, if any, could have done this better than he did.

In the final decade of his life, Nansen devoted himself primarily to the League of Nations, following his appointment in 1921 as the League's High Commissioner for Refugees. In 1922 he was awarded the Nobel Peace Prize for his work to help displaced victims of World War I and consecutive conflicts. Among his initiatives was the introduction of the 'Nansen passport' for stateless persons, which was an internationally recognised document, and the start of the then United Nations Organization's quota system for refugees, which still exists today. He worked on behalf of refugees until his sudden death in 1930, after which the League of Nations established the Nansen International Office for Refugees to ensure that his work continued. Fridtjof Nansen was one of the greatest multitalented persons that Norway has ever fostered.

In the harmonic portrait to the right, dominated by the peasant-blue colour of traditional Norwegian countryside furniture, Erik Werenskiold depicted some of Nansen's characteristics: an introvert, a dreamer, and mild-tempered. The portrait is currently on display at Polhøgda, Nansen's home in Oslo. If we compare the portrait of Nansen with that of Amundsen, the pronounced difference between the two polar heroes is very readily apparent. Both men had unique personalities, but only one of them was a true friend of other people. Nansen had a unique, multitalented, and multifaceted personality. There has only ever been one Nansen. ∎

Scientists: Gunnar Isachsen

Gunnerius ('Gunnar') Ingvald Isachsen (1868–1939) was a Norwegian military officer and polar scientist. In 1888 he entered the Norwegian Military Academy, where he took courses at the Marine Observatory in Wilhelmshaven and engaged in marine research in Bergen. From 1898 to 1902, Isachsen was topographer on Otto Sverdrup's Fram expedition to the Arctic archipelago, i.e. north-eastern Canada. During the voyage he mapped exceptionally large areas of hitherto unknown islands in northern Canada, mainly by undertaking long journeys by sledge. The aforementioned places included Ellef Ringnes Island and King Christian Island. The now abandoned remote Arctic research station named *Isachsen*, on the western shore of Ellef Ringnes Island in the Sverdrup Islands, was named in his honour. Few know that the second Norwegian Fram expedition mapped much of the Canadian archipelago – a fine contribution to another Arctic country that has so much in common with Norway. History could have turned out very differently were it not for the fact that after Sverdrup had annexed the no-man's-land for Norway, the Swedish–Norwegian Government did not follow up the claim.

Between 1903 and 1905, Isachsen was a member of the French military service in Algeria and Paris. This French intermezzo apparently drew his attention to French and Monégasque oceanographic research. During the period 1906–1910 Isachsen entered the realm of Svalbard and led topography and bathymetry research expeditions in Svalbard. Two of these expeditions were financed by Prince Albert I of Monaco. Isachsen made sledge trips over many glaciers to conduct his topographical studies, mapping 1930 square kilometres, covering *Smeerenburgbreen*, *Lilliehökbreen*, *Monacobreen*, *Isachsenfonna*, and *Fjortende Julibreen*. Isachsen led his own government-financed expeditions to Spitsbergen in 1909 and 1910. As a result of these expeditions, he founded Norway's systematic research work in Svalbard. In 1914, tragically, a fire destroyed his house, including all his records and original maps. Much of Isachsen's work though, is published in French (e.g. *Exploration du Nord-Ouest du Spitzberg, entreprise sous les auspices S.A.S. le Prince de Monaco par la Mission Isachsen*). In order to gain a good impression of the dimension of Isachsen's work, one has to visit the Musée océanographique de Monaco, where *La Mission Isachsen* is a major part of the exhibition regarding the Arctic, including the several square metres large oil painting *La mission Isachsen engagé sur un glacier franchit un port de glace*, by Louis Tinayre (page 221).

Isachsen served as a regular salaried officer until 1917 and as the Norwegian Government's technical delegate both to the signing of the Svalbard Treaty of Paris in 1914 and to the Paris Peace Conference in 1919. Towards the end of his life he was the government's whaling inspector in the Southern Ocean in the period 1929–1930, and the leader of the fourth *Norvegia* expedition, which circumnavigated the South Pole in the years 1930–1931.

The portrait of Gunnar Isachsen (left) was painted by Louis Tinayre (pages 11, 166, 221, and 260–261) and is on display at the Norwegian Polar Institute, in Tromsø. Isachsen must have just returned from one of his many sledge trips over glaciers, as he is shown unshaved, exhausted, and sporting snow goggles. The portrait suggests that, once again, Isachsen had just given his very best for the benefit of his research.

In the photograph (right), Isaksen studies some papers onboard of *Princesse Alice II*. The Norwegian support ship of the expedition, *Kvedfjord*, and a proud crew (left)

Scientists: Adolf Hoel

Adolf Hoel (1879-1964) was an influential Norwegian geologist and polar researcher, probably one of the greatest that Norway has ever fostered. His methods did not comprise spectacular expeditions or adventures to reach the North Pole. Rather, he used all his time to carry out, promote, and administer Norwegian exploration and presence in polar regions, and mainly Svalbard.

During the period 1907-1945, Hoel worked as a geologist and was a leading Svalbard researcher. In 1907 he joined the cartographer Gunnar Isachsen on his epoch-making expeditions to north-western Spitsbergen. Thereafter he spent every summer in Svalbard until Norwegian sovereignty was established in 1925. Hoel had a significant role in annexing coalfields for Norway. In addition, he played an active role during the founding of the Store Norske Spitsbergen Kullkompani, the leading coalmining company in Svalbard. After World War I, Norway provided evidence of systematic work in Svalbard based upon Hoel's and Isachsen's expeditions. As a consequence, Norwegian sovereignty over Svalbard was internationally accepted.

Hoel also played an essential role in securing Norway's claim to sovereignty over Dronning Maud Land in Antarctica. Further, he was one of the initiators of the unlawful Norwegian occupation of Eirik Raudes Land in East Greenland (1931-1933). In 1928 he established Norges Svalbard- og Ishavsundersøkelser (Norways Svalbard and Polar Sea investigations), a forerunner of today's Norwegian Polar Institute, which he managed for 17 years. During much of this time Hoel was also a professor, pro-dean, and dean of the University of Oslo.

Hoel's active support of the Nasjonal Samling (the Norwegian Fascist Party) and support of Norway's Vidkun Quisling regime during World War II resulted in prosecution. He was imprisoned for about one year between 1945 and 1946. Later, after he was accused of high treason in 1949, he was sentenced to 18 months imprisonment. He subsequently lost all right to his positions and medals. Thus, he tragically disappeared from the limelight of Norwegian Arctic research. He died in 1964 and it can be taken for granted that his last decade was not one of his best. Without Hoel, Norway's position in Svalbard would have been much weaker. Despite his political views and support of Quisling, Norway owes him a great deal – nothing (and no one) is ever so bad that it is not good for something, and this holds true for Adolf Hoel. Even traitors can play an important role for a free and democratic country.

The painting to the left shows a serious, fierce, but self-assured looking Hoel, gazing through his conspicuous glasses towards Kongsfjorden from the beach of Ny-Ålesund. Behind Hoel, the spectacular vista of the characteristic mountains Tre Kroner, in the inner part of Kongsfjorden, can be discerned. The painting was done by the painter Gunnar Wefring in 1936, when he spent a few months in Svalbard. On page 221, Louis Tinayre's painting from 1907 shows Gunnar Isachsen's scientific activities, which Hoel's work was based upon – the cartography of western Spitsbergen. Clearly, treacherous glaciers had to be crossed in the process of mapping the area. ∎

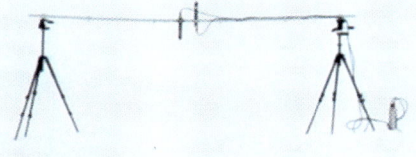

Hunters:
the Russian Ivan Starostin

Little is known about the Pomor hunters that regularly passed through Svalbard and left huts, graveyards, and even ships behind. The Russians came and went, along with so many other nations that frequented Svalbard in pursuit of their economic and political interests: the Dutch, English, Americans, Russians, Norwegians, and others. A multitude of orthodox crosses and graves (see page 66) tell the tale of Russian engagement and the hardship, sickness, and misery that the hunters must have endured.

The regions used by Russian hunters were extensive and almost covered the entire central and western archipelago. Most of the Pomor hunters came for one season, but some stayed for several seasons. Russians hunters were once quite active in Russekeila, in the outer part of Isfjorden. Russekeila was one of the permanent Russian overwintering sites of those times, and the remains of several huts and many burial sites can still be found there. The bravest Pomor of all to have overwintered in Svalbard was Ivan Starostin, who lived and died here. He was known as the 'King of Svalbard'.

Ivan Starostin first started to travel to Svalbard in 1780, when he was sent by the monks of the famous and prosperous Solowetsky monastery in the White Sea. It is said that he spent 39 winters in Svalbard, 15 of them consecutively, which was (and is) an impressive record. Starostin died in 1826 and was buried on a cape named after him: Kapp Starostin. The king is dead! Long live the King of Svalbard! ■

Hunters: the Norwegian Hilmar Nøis

The hunting place Fredheim, which was the home of the hunter Hilmar Nøis (1891–1975), is located in Sassenfjorden, part of Isfjorden. Nøis overwintered in Svalbard 38 times, but Fredheim was not his only base. There are just two villas in Svalbard: Villa Amundsen in Ny-Ålesund and the main house of the 'King of Sassen', Villa Fredheim. The latter building was started in 1924, but rebuilt and widened several times. It is one of the largest hunting stations in Svalbard. Two women that also spent winters in Svalbard, Ellen Dorthe Johansen Nøis and later Helfrid Nøis, contributed much to the quality of the house. Both of them changed the cabins into a comfortable home, with a flagpole, potted plants, curtains, and tablecloths. The base was a far cry from the functionality of contemporary cabins that were entirely dedicated to survival and hunting. However, Fredheim also had side stations, encircled by self-firing traps for polar bears and traps for Arctic foxes.

To the east of Fredheim lies Gammelhytta and Danielbua, which were built by Hilmar Nøis's father, Daniel Nøis. The present-day governor of Svalbard conducts searches for many of the surviving huts and today Daniel Nøis's two huts appear as they did in the heyday of hunting. They have dried moss built into the walls for insulation, birch bark on the walls and roofs, and piles of peat heaped against the walls and on the roof.

Hilmar Nøis is often referred to as the King of Sassen. Obviously, the only kings in Svalbard are hunters. Nøis was only 18 years of age when he overwintered in Svalbard for the first time. He was 72 years old when he overwintered for the last time, together with his wife Helfrid. Caught by the freedom, the landscape, and nature in Svalbard, hunting became a profession, a lifestyle for Hilmar, not a job. His profession was characterised by working all day long and being constantly exposed to risks and, at times, to extremely low temperatures.

Hilmar Nøis addressed the core of his fabled lifestyle, which still causes many a Norwegian soul to oscillate, with the following words: 1 will always remember the freedom on the wide expanse and the mountain plateaus.' No one in Norway's High North could compete with the aura of fame that wrapped the hunters of Svalbard.

The photograph shows a deeply satisfied, smiling Nøis in front of Villa Fredheim. His satisfaction derives from a life that was developed according to his innermost wishes. This extraordinary satisfaction is well manifested in the vivid eyes of Nøis in the photograph. ■

Women on Svalbard: Léonie d'Aunet

There have been false rumours about brothels in Smeerenburg during the heyday of whaling, but the Parisian Léonie d'Aunet (1820–1879) was probably the first women to visit Svalbard, when she was only 18 years of age. She was not a scientist, but the fiancé of the La Recherche research expedition painter François-Auguste Biard. She travelled on land in men's clothing to Hammerfest in northern Norway in order to join the La Recherche expedition in 1839. At the time, it was forbidden for women to join a voyage on board on a French Navy vessel. Furthermore, it was generally considered inappropriate to have ladies on board any ship, not least because it was believed that they would bring misfortune to the ship and its crew. In 1854 d'Aunet published the book *Voyage d'une femme au Spitzberg*, in which she described the trip and expedition. Her book contains details of the famous La Recherche expedition that are not mentioned in scientific publications. With the coquetry, naivety, and snobbishness of a young women from the 'better' circles of Paris her views and prejudices drizzle highhandedly over the seemingly less advanced cultures and simple people that she observed and encountered en route to Hammerfest.

The expedition visited Magdalenefjorden and, while there, d'Aunet described a few essential scenes. Considerably impressed by the Arctic landscape after long periods at sea, her writing develops into an emotional crescendo: 'These polar ices, which no dust has ever stained, as spotless now as on the first day of the creation, are tinted with the most vivid colours, so that they look like rocks composed of precious stones: the glitter of the diamond, the dazzling hues of the sapphire and the emerald, blend in an unknown and marvellous substance.' At the entrance of the fjord, d'Aunet visited the burial site Gravnesset, which Biard immortalised in his famous painting Magdalena Bay (see page 250). There, with ice, coldness, and potential death all around her, her attraction for the polar world faints and her desire for life in more civilised circumstances surfaces: 'In the middle of this work of burial, I was seized with an indescribable horror; the thought came upon me that I was doomed, perhaps, to lay my bones among these dismembered skeletons. I had been forewarned of the perils of our expedition. I had accepted the warning and fancied that I comprehended all the hazard; yet these tombs made me for the moment shudder, and for the first time I dwelt with regret on the memories of France, my family, my friends, the blue sky, the gentle and serene life which I had quitted in order to incur the risks of so dangerous a voyage.'

Léonie d'Aunet returned to Paris in safety. Satisfied with her Arctic encounters, she did not attempt a second expedition. She married Biard (page 250), who painted a touching portrait of his young wife in 1842, which is now on display at the Châteaux de Versailles et de Trianon, in France (page 227). The happy years spent by the couple did not last long, as none less than the great Victor Hugo fell in love with Leonie in 1844 and she became his mistress. Such behaviour was not unusual at that time. Their rendezvous did not remain private for very long and Biard was not amused. Detectives had found Hugo and Léonie committing adultery, and Leonie was sent to jail for a couple of months while Hugo left the police station a free man because his fame rendered him immune to prosecution in Paris of those days. After a difficult divorce, d'Aunet turned to writing and became a well-known author, particularly for her diary of her journey to Svalbard.

History does not inform about whether d'Aunet resumed her relationship with Hugo or about any other partners in the remaining 35 years of her life. However, compared to the ideal woman of the time, she had a refreshingly independent and self-confident lifestyle. Perhaps the viewer can detect in Biard's empathic portrait (right) that already at young age d´Aunet felt at odds with her time? Do her eyes tell stories that have yet to be realised or of controversial facets of her life? She appears to be in the middle of a dialogue between her inner self and the outer world. ■

MESMERIZING EXTRACTS FROM LÉONIE D'AUNETS VOYAGE D'UNE FEMME AU SPITZBERG

Léonie d'Aunet's description of her visit to Magdalenefjorden is truly inspiring reading and worthy of the presentation of a few extracts here. With a young person's élan and overwhelmed by the sight of an Arctic fjord, calving glaciers, and the ancient graveyard of Gravnesset, d'Aunet ran out of words and searched instead for analogies and elucidating vocabulary. Such was life, when, with an open mind and a full soul, the young woman had attempted to put onto paper what must have seemed utterly unimaginable, given that she lived far removed from nature, in remote Paris.

Floating islands, incessantly undermined by the sea, change their outline every moment; by an abrupt movement the base becomes the summit; a spire transforms itself into a mushroom; a column broadens out into a vast flat table, a tower is changed into a flight of steps; and all so rapidly and unexpectedly that, in spite of oneself, one dreams that some supernatural will preside over those sudden transformations. At the first glance I could not help thinking that I saw before me a city of the elves, destroyed at a blow by a superior power, and condemned to disappear without leaving a trace of its existence. Around me hustled fragments of the architecture of all periods and every style: campaniles, columns, minarets, pyramids, turrets, cupolas, crenulations, volutes, arcades, façades, colossal foundations, sculptures as delicate as those which festoon the shapely pillars of our cathedrals – all were massed together and confused in a common disaster. An ensemble so strange, so marvellous, that the artist's brush is unable to reproduce, and the writer's words fail adequately to describe!

This region, where everything is cold and inert, has been represented, has it not? as enveloped in a deep and sublime silence. But the reader must please to receive a very different impression; nothing can give any fit to the idea of the tremendous tumult of a day of thaw at Spitzbergen. The sea, bristling with jagged sheets of ice, clangs and clatters noisily; the lofty littoral peaks glide down to the shore, fall away, and plunge into the gulf of waters with an awful crash. The mountains are torn and splintered; the waves dash furiously against the granite capes; the icebergs, as they shiver into pieces, give vent to sharp reports like the rattle of musketry; the wind with a hoarse roar, scatters tornadoes of snow abroad. It is terrible, it is magnificent; one seems to hear the chorus of the abysses of the old world preluding a new chaos.

Never before has one seen or heard anything comparable to that which one sees and hears there; one has conceived nothing like it, even not in one's dreams! It belongs at once to the fantastic and to the real: it disconcerts the memory, dazes the mind, and fills it with an indescribable sense of awe and admiration. But if the spectacle of the bay had something magical in it, ominous and gloomy was the scene on shore. In all directions the ground was white with the bones of seals and walruses, left there by the Norwegian or Russian fishermen, who formerly visited these high latitudes for the purpose of collecting oil; for some years, however, they have abandoned a pursuit which was much more dangerous than profitable. These great bones, bleached by time and preserved intact by the frost, seemed so many skeletons of giants – the past dwellers in a city that had finally been swallowed up by the sea.

I found myself very speedily in the middle of a cemetery; but this time, the remains lying on the frozen snow were human. Several coffins, half open and empty, had formerly been occupied by human bodies, which the teeth of the polar bear had recently profaned. As, owing to the thickness of the ice, it is impossible to dig graves, a number of enormous stones had, in primitive fashion, been heaped over the coffin lids, so as to form a defence against the attacks of wild beasts; but the stout limbs of the polar bears had removed the stones and devastated the tombs; a throng of bones strewed the shore, half broken and gnawed the pitiful remains of the bears' banquet. I carefully collected them, and replaced them piously in their proper (place). ■

Translation: W. H. Davenport Adams (1905)

Women on Svalbard: Hanna Resvoll-Holmsen

Hanna Marie Resvoll-Holmsen (1873–1943) was a Norwegian botanist and nature conservation activist. She was also an active photographer and wrote lyrics. She was among some of the first students to be awarded a Master of Science degree in botany in Norway. However, earlier still, her sister Thekla had defended her master thesis in 1899. Obviously, both women came from an open-minded and progressive family that cared for their daughters' higher education. Hanna's youth was characterised by sickness. Later, after an unhappy marriage, when she was in her thirties, she started to study. Hanna Resvoll-Dieset became a pioneer in biological investigations in Svalbard: she was one of the first Norwegians and the first female scientist to work there. In 1907 she went to Svalbard to carry out fieldwork for her thesis. Her interest in alpine plants first led her to the archipelago. She participated as a botanist during Prince Albert I of Monaco's expeditions to Svalbard, onboard the research vessel the *Princesse Alice II* and the motor cutter *Kvedfjord* (page 219). The leader of the expedition, Gunnar Isachsen, originally had not made any plans to conduct botanical investigations, but allowed Resvoll-Dieset to participate after her study colleague Adolf Hoel, had recommended her. She was set ashore, alone, with a gun, a camera, a plant press, adequate provisions, and what she described as 'my airy home', her tent. Wearing a shirt of appropriate length, a fine-looking jacket, and a handsome hat, she must have represented the very height of Svalbard's haut couture at the start of the 20th century! Weeks later, Svalbard's best-dressed biologist was picked up again, still impeccably clothed after weeks in rough nature and living in a tent.

In 1908 Resvoll-Dieset returned to Svalbard, with her own expedition. She went there together with the first Norwegian Svalbard Expedition, led by Adolf Hoel. She sampled plants and investigated the environmental conditions for growth. During her expedition she acquired a good overview of Svalbard's vegetation along the west coast and adjacent fjords. Some legs of the 1908 expedition were carried out in the company of Gunnar Holmsen and Hjalmar Johansen. One year later, Resvoll-Dieset divorced and married Holmsen, a state geologist. Thus, Svalbard life deals with many aspects, including finding the right partner. Less hospital circumstances and a spectacular and demanding nature clearly do not prevent humans from uncovering essential facets of their personal lives. May be, on the contrary, the isolation provides insights into our innermost souls that the veneer of our mundane culture precludes.

Resvoll-Holmsen photographed vegetation and landscapes and was probably the first biologist to use colour photography as a means of documentation. Based upon her fieldwork in Svalbard, she defended her thesis in 1910, which, incidentally, was written in French, probably in order to have it published in 1913 with the title *Observations botanique*, as part of Albert I extensive works in Svalbard, *Exploration du Nord-Ouest du Spitsbergen: entreprise sous les auspices de S.A.S. le Prince Albert de Monaco par la Mission Isachsen*. Already in 1927 the first edition of the popular *Svalbard flora* (at the start authored by Resvoll-Holmsen) was published. The core of the book is still based upon Resvoll-Holmsen's work. In 1910 she was personally invited by the Prince to attend the opening of the splendid and spectacular Musée océanographique de Monaco, partly as a journalist for the Norwegian newspaper *Aftenposten*. Part of the exhibition from days dealing with Prince Albert I's expeditions to Svalbard can be studied in the museum today, particularly Louis Tinayre's brilliant paintings of Svalbard glaciers.

Resvoll-Holmsen became docent in plant geography at the University of Oslo in 1921 and in 1937 became an elected member of the Norwegian Academy of Science. She was one of the first Norwegians to pay attention to nature conservation and her influence on the Svalbard Treaty is evident in § 2, which mentions that Norway has the right to introduce rules that secure or reinstall plants and animal life in Svalbard's terrestrial and marine regions. In concert with Adolf Hoel, Resvoll-Holmsen argued the case for protected regions, and already by 1932 the protection of major regions of botanical interest had been established.

Hanna Resvoll-Holmsen was a unique character in her time. She was seemingly fully absorbed in bourgeois life, yet at the same time she developed her interests and career independently, in a manner characteristic of women today. She deserves far more recognition for her role in Arctic research and nature conservation. However, perhaps we should also pay due attention to Gunnar Holmsen, who, at odds with his time, must have supported his Hanna with all his emancipated strength? ■

Women on Svalbard: Wanny Woldstad

Wanny Woldstad (née Ingebritsen) (1895–1959) was the first female hunter in Svalbard. At the age of 15 years she moved from the family's seaside home to Tromsø in order to receive more education. At 20 years of age she married Othar Jacobsen, a shopkeeper. She gave birth to two sons, Alf and Bjørvik. Her husband died in the Spanish flu epidemic in 1918, only 27 years old. Wanny Woldstad secured a position at the prestigious Grand Hotel in Tromsø and married a baker, a local man named Martin Mikal Woldstad (1861–1939), who was 34 years her senior.

Wanny Woldstad was a clever and independent woman, a quality that was not rare among females along the coast of northern Norway, who were running a household, farm, and children during lengthy periods of time when men were away fishing, either as crew for Norway's big fishing fleets or on smaller vessels in the Arctic. In some cases, the men did not return from their dangerous occupations, and women were forced to take on sole responsibility for their family's livelihood.

Wolstad first acquired a driver's licence and then a taxi driver's licence and thereafter worked as a taxi driver in Tromsø using her own car. At times, she had a higher income than her husband. Through this work she met hunters that had overwintered in Svalbard and she became fascinated by their stories and life in the wilderness. Such fascination with hunting life in the wilderness is deeply rooted in the High North, where elements of man's early civilisation seem to linger in the depths of many souls. To be free and to have to rely solely on oneself seems to induce an irresistible attraction. I have met many that have felt the 'call of the wild'. Among these is my father-in-law Rolf Moelv, who before World War II shared a room at school with Bjørvik Jacobsen (who later became my neighbour) and was invited to join Wanny Woldstad when she planned a new overwintering in Svalbard, despite the fact that Rolf's mother was not too fond of the idea!

During the winter of 1932–1933 and the following four winters, Woldstad overwintered at Hyttevika in Hornsund, together with the hunter Anders Sæterdal. It could be assumed that Sæterdal and Woldstad were a couple, but in her writings she refers to him by his full name but as her 'partner'. Her young sons joined her the winters of 1933–1934 and 1934–1935. It is told that she baked 900 loaves of bread and countless cakes during one winter in Svalbard. She had also responsibility for keeping the house clean, for washing and repairing clothes, and mending ropes and nets. In addition, she was fully active during the hunting season. After World War II, she moved from Tromsø and took a job as a housemother, but gave presentations about her adventures in the wilderness. In 1956 she published the book *Første kvinne som fangsmann på Svalbard* (First Female Hunter in Svalbard), which resulted in further invitations to give talks. She died tragically, after a truck hit her during a visit to her son Bjørvik in Sørkjosen. She was about to publish a second book: *Kvinnen fra havgapet* (Woman from the Edge of the Sea). The 'edge of the sea' is a heritage site for many Norwegians.

In the introduction to her first book, the Arctic celebrity Helge Ingstad wrote: 'She was a peculiarity in a field that had been men's domain since the Stone Age'. The words are patronising and conservative, yet they show full respect for her work. Moreover, there always have been women, if not in Svalbard, then along the north-Norwegian coast, that have been used to living in a setting where a job was a job, irrespective of their gender. Wanny Woldstad, who had a small statue and was light humoured and easy to get along with, combined the traditionally separate roles of mother and father, lover and colleague, indoor and outdoor work, and man and woman into one entity, seemingly with ease as the most natural thing on earth. My impression is that she was comfortable in all of these roles that were, after all, her life.

Since Helge Ingstad wrote his introduction, the Stone Age has definitely come to an end. Wanny Woldstad, in common with many other women, listened to her inner voice. However, she did not only listen, but was brave enough to realise aspects of her life that were not common in her days. This may be easier to do now than previously, but it is nevertheless an art to become what is inside us, and not tread on the toes of our compatriots in the process.

On page 233, a proud Wanny is shown in front of two recently killed polar bears, perhaps her first successful hunt. ∎

Miners

In contrast with explorers, scientists, and hunters (who were often characterised by their specific, outspoken personalities), miners did not appear into the public eye as individuals. Few, if any, essays, books, or historical reports exist on individual miners, and this absence reflects the nature of mining. Regardless of the relevance for society and the livelihoods of families that depend upon the work done by miners, mining is a strongly collective profession, not an individualistic one. There is never a hero miner (except in Soviet propaganda). Rather, there is always the successful team, the professional assembly, and the brave comradeship. Miners have been present in Svalbard during the course of the last 120 years, and for most of this time mining has represented the economic backbone of the archipelago. Despite, the significance of mining, there does not appear to be any description of individual miners. Instead, they are treated as a group, as elsewhere in the world.

Undoubtedly, life as a miner in Svalbard has always been tough. Throughout the 20th century, Svalbard's mining operations were dogged by accidents. In common with other mining districts in the world, the local populations were constantly in fear of lethal accidents occurring. Hardly a year went by without at least 1 fatality. For example one day in 1952, two separate gas explosions killed 6 and 9 miners respectively, and 1 year later 19 miners were killed in Ny-Ålesund. No fewer than 80 miners died during less than 50 years of mining in the small community of Ny-Ålesund.

When Liv Balstad left her high-society circles in Oslo in 1946, having just married the governor of Svalbard, she was the first influential woman to experience life on Spitsbergen. Ten years later, her memoirs created a political stir in Norway, with their revelations of the harsh living and working conditions in the forgotten outpost. The memoirs contained little information about the Soviet Union's mines, but perhaps their situation was no better as compared to the Norwegian mines? As recently as 1997, 23 miners died in a mine fire in Barentsburg. An even greater tragedy had happened one year earlier, in 1996, when a Russian plane carrying 141 people (including miners' families) crashed close to Adventdalen. Recently, a small chapel was erected in the centre of Barentsburg to commemorate the lives of those lost in both tragedies.

Socialist states show great appreciation for their workers, particularly those that worked in heavy industries and construction work, or as soldiers, pilots, and even miners. In honour of all Svalbard's miners, we show three of them here, proud and satisfied, in impeccable social-realistic style, and probably from Pyramiden. The painting is by the Russian artist A. Strakhov and is currently on display at the Barentsburg Museum. ■

People: now

Svalbard Life is not only about the past, but also very much an account of what life prevails there at present and what may emerge in the future. Among the most appreciated elements in any system are its people. Among the large number of interesting individuals that make Svalbard such an enchanting and lively place, we wish to present a few. The selection is by no means representative of the entire population, but the reader should gain an impression of some of the individuals in the archipelago. The assortment gives 'the people' individual expressions. As Svalbard is not a lifecycle society, I asked the interviewees about where they came from, what they did, what their motivation was for working in Svalbard, and what aspirations they had for their future in the Arctic.

The photograph depicts the new and old Longyearbyen, represented by the modern Sysselmann (District Governor's) office in the foreground and the heart of the coal transportation cable line in the background. ■

ODD OLSEN INGERØ, LONGYEARBYEN

The current governor (*sysselmann*) of Svalbard is Odd Olsen Ingerø, who is the highest representative of the Norwegian state in the archipelago. The position of district governor can be likened to the combined role of county governor and chief constable on the Norwegian mainland. I meet him in the governor's office, a spectacular, modern building that not only overlooks Longyearbyen and the Adventfjorden, but also seems, mildly but decisively, to command respect for the law. Olsen Ingerø has held the position of governor twice, first between 2001 and 2005, and since 2009. Under Norwegian law he can hold the position until 2014. When I asked about when his interest in Svalbard started, he replied with natural eloquence: 'When I was a boy, a new priest came to my village [in Norway], who had been in Svalbard for nine years. He gave talks, presented pictures, and told stories that ignited my interest.'

Initially, Ingerø's childhood experience resulted in two decades of employment in the small north-Norwegian town of Kirkenes, close to the Russian border, where he was a chief constable for 15 years. During his first term of office as governor of Svalbard, Olsen Ingerø was joined by his family, but they did not accompany him when he started his second term of office. Currently, he lives alone in Svalbard, and is not an exception in this respect as many people who work in Svalbard commute to the mainland. When jokingly, I asked the rhetorical question 'I presume you like your job?', he replied:

Yes, indeed! This is a far more versatile profession than any other on the mainland. The Svalbard community is a special one. On the one hand, everybody has a job and a good income, the population is generally healthy, nobody is born here, and nobody in need of nursing can retire here. On the other hand, one has very close contacts to the entire population. In a manner, we all know each other well and we live life very much in the public eye. This makes the Svalbard society special.

Olsen Ingerø added:

Knowing so many people so well involves having a lot of highly appreciated qualities, but also encompasses challenges. My official duties imply strict independence of whom I know and meet, and my experience from a small settlement such as Kirkenes helps me to separate between my official and private roles. I continuously have to keep my role in mind and adapt to the fact that I never can be a very private person in public.

Where can we learn the skills of combining one's official role and private role simultaneously and all the time remain relaxed and be 'ourselves'? Such skills are not taught anywhere and this is an art that the governor of Svalbard has to master. I appreciate such professionalism. It provides substantial quality to the Svalbard society: one is seen, one is treated with due respect and one is taken care of, but in return one has to contribute equally. ∎

LENA ROMANENKO, BARENTSBURG

I met Lena, a slender, charming woman, in the shop at the Barentsburg Museum. She proudly welcomed me and spoke in fluent English. In Barentsburg, she is currently an important face for visitors from the outside world. She probably meets all visitors to Barentsburg and provides them with excellent service and an open mind. She came to Barentsburg in 2003 and has two jobs. She is the leader of the local sports club and works at the museum. However, she pointed out 'But my hobby is dancing. I am the leader of the local dancing group.' Indeed, I have seen her on stage in action with her group, as part of an entertainment show for visiting tourists. During the show I asked myself whether any other settlement the size of Barentsburg could provide a 1-hour long potpourri of traditional songs, including a dance performance? The achievement was impressive. Lena Romanenko came to Barentsburg from Ukraine, together with her husband, who works in the mine, but Lena is Russian. They left their children, of school age, with their grandparents in Ukraine. I understand that one of the reasons why the couple works in Barentsburg is that the income is better there compared to their home country. They have bought a house at home and need financial resources to make it modern and pleasant. In addition, they need money to pay for the education of their children, which is a new challenge facing many families from former USSR states. The best education is no longer provided free of charge. Lena Romanenko smiled and said: 'A few more years will be enough up here and then we go home.' Svalbard has functioned in this manner for many employees: work and earn as much as possible, save money, support one's family at home, and get out of an economic bottleneck. ■

ROGER JACOBSEN, NY-ÅLESUND

Roger Jacobsen is currently the former director of Kings Bay A/S, which runs the Ny-Ålesund research settlement. I arrived on R/V *Helmer Hanssen*, together with an entire team of outreach cruise participants and artists, all of whom were eager to go ashore. I met Roger Jacobsen at the Ny-Ålesund dockside. He had organised a presentation of the various institutions that are gathered in Ny-Ålesund. It was summer, the sun was shining brightly, and the glaciated vistas of Kongsfjorden were welcoming on the horizon. With disregard for the calendar, we were dressed in a manner that would be appropriate for an average winter day in central Europe. Jacobsen smiled; after all, it *was* summer! He has 27 employees that facilitate research in the world's northernmost settlement. In summer there are up to 180 persons in the settlement, but when autumn approaches and migrating birds come to their senses and fly towards the south, less and less research is carried out. In winter, the number of permanent staff decreases to 35. I asked Roger, 'Why did you come to Svalbard?' and he responded, 'I have always been attracted to nature and exiting assignments. Once you have got the idea of Svalbard on your brain then you easily look for chances to realise this idea. In Ny-Ålesund I can combine an exiting job with work that is very relevant for society.' The position as director is usually held for three years, with the possibility of prolonging it by a further two. In response to my question 'Do you foresee your future in Svalbard or on the mainland?', Roger replied: 'I am on leave from a job on the mainland and will return to it when my contract in Kings Bay is over'. Then I asked: 'Will you return to Svalbard again in the future?' He informed, 'That is not impossible'. I detected hope in Jacobsen's voice, when he continued to say 'I will need a few years to think about this.' I am convinced that I will find the answer to my question if, and when, Jacobsen starts his next term of office. We have both our families with us in the same village in northern Norway, and meet, along with the commuters, every Monday morning at the local airport. The world in the High North is small, and only the physical distances are vast. ■

HANS ROAR HANSEN, TROMSØ

Few ships sail to Svalbard all year round, and these are mainly fishing and cargo vessels. Among the many

people that make a living from transport and marine resources in the Svalbard region, I talked to Hans Roar Hansen. This was an obvious choice because he is a well-known figure among all those engaged in research in the Barents Sea and the waters adjacent to Svalbard. He is one of two captains of the University of Tromsø research vessel the *Helmer Hanssen* (formerly the *Jan Mayen*). I met him on the bridge of the research vessel, off the coast of Svalbard. Hansen explained that R/V *Helmer Hanssen* spends about 5 months each year in the waters around Svalbard, in order for those onboard to carry out research and educate students. Over the years Hansen has sailed around the entire archipelago many times, but there are still a few places that he has not yet seen. This response was typical of a modest, understated comment made by a Norwegian, as few can have seen as much of Svalbard as the crew of the *Helmer Hanssen*. When I asked, 'You work up here so often, do you live in Longyearbyen?', he replied 'No, I do not live there. Actually, I have my family in the very south of Norway, but I was born and raised in Tromsø, where also this ship has its home port'. Then, I asked 'And will you continue to work around Svalbard'?, and his response was 'As long as I work on this ship, I will be here several times each year', before turning his attention back to the vessel's instruments and the next instrument that was to be lowered into the sea. What could oceanographers hope to achieve at sea without the dedicated fulfilment of duties of a research vessel's crew? A good crew takes pride in making real the crazy investigations of scientists, a precondition of scientific success. It is said that behind a great man stands a great woman, but behind a good oceanographer there often stands the dedicated crew of a research vessel. In this respect, the captain is essential for setting the standards, the attitude, and the atmosphere for all those onboard. On behalf of the scientists working on the R/V the *Helmer Hanssen* we offer thanks to such captains and their crews for providing efficient working conditions and a pleasant atmosphere on board ship. ∎

FIONA DANKS, NY-ÅLESUND/TROMSØ

The Arctic is known for its mobile migrants that perpetually connect these remote regions with more southerly ones. Svalbard is no exception. I met Fiona in early summer in Ny-Ålesund. The snow still covered the ground, but a research group was looking for freshwater run-off in order to carry out an experiment on the effect of rivers on marine organisms. They wondered whether our colleagues at the Sverdrup Station in Ny-Ålesund could advise them on where to go. The answer was a resounding yes, Fiona could. She is a terrestrial ecologist and a geographer. Currently, she is the manager of the modern and comfortable Sverdrup Station run by the Norwegian Polar Institute. I asked her 'How long have you been here in Ny-Ålesund?' and she replied 'Two years by now, on and off. I commute between Tromsø and Ny-Ålesund'. I was curious to know what had led her to northern Norway and Svalbard, and she responded: 'I almost made my way around the entire Arctic in the last 15 years, starting in Yukon in Canada, then moving to Alaska and Greenland and finally to northern Norway and Svalbard'. When I asked, 'Are you infected by the Arctic bug?', Fiona laughed and nodded affirmatively: 'The Arctic bug can shed light on my selection of places!' She strongly wishes to continue her work and, in common with the sentiments of many other scientists, she added 'I could not think of anything more interesting than what I am doing ... I would like to continue with this work for the rest of my life, but commuting back and forth between Tromsø and Ny-Ålesund is a little disruptive. I am looking for a better balance'. Clearly, balance is something that all of us struggle with. In Schubert's *Lied* (song) cycle *Winterreise*, the deeply troubled and heartbroken wanderer shouts into the woods 'Happiness, where are you?', and the echo responds 'Not where you are!' ∎

CARLOS DUARTE, MALLORCA

Among the many migrants to the Arctic, I met Professor Carlos Duarte, a researcher onboard the research vessel the *Helmer Hanssen*. Duarte has been deeply involved in a series of investigations for the European Union research project Arctic Tipping Points. He was leading a Spanish team from his research organisation Mediterranean Institute for Advanced Studies, in Mallorca. I wondered why an entire Spanish team would want to go conduct research in Svalbard and the Arctic Ocean instead of studying in the more pleasant and comfortable Mediterranean region? Duarte's response was: 'We have global research interests, also in polar research. Spitsbergen is the most convenient platform to carry out research in the European sector of the Arctic Ocean. Since 2006 we have been conducting research in the Arctic and in 2007 Spain brought its most prominent research vessel the *Hesperides* to Svalbard and beyond'. I then asked 'How many people from Spain are involved in Arctic research', to which he responded 'The teams are of variable size, but during major cruises we bring 20 to 37 scientists and technicians ... Not only on board'. He added, 'We also run experiments on land in Longyearbyen. We are mostly interested in biogeochemical cycling and marine ecology in the Arctic. The significance of the Arctic Ocean for the entire globe keeps our interest in the region active'. Indeed, in the times of the 'Anthropocene' (the era in which human activity influences the biogeochemical cycling of the entire earth) the whole globe has become a research field, and what happens in the Arctic is also of direct relevance to more southern countries.. ∎

TOVE GABRIELSEN, LONGYEARBYEN

The education of a future generation that knows about life in the midst of 'Anthropocene' is essential for our joint future. This puts teachers and supervisors in the limelight. I meet Tove in bright sunshine outside a unique building, a landmark in Longyearbyen: the University Centre of Svalbard (UNIS). Tove Gabrielsen is a marine biologist and currently holds a position at the Department for Arctic Biology at UNIS. Her field of interest includes the smallest organisms that are carried by moving water. She left the University of Oslo four years ago to work in Svalbard. She did not come alone, but accompanied by her family. I asked how she manages a full-time research-, a teaching job, and research ambitions with two children, aged 5 years and 9 years respectively, and a husband that also needs an attractive job? Her answer was typical of many people that have moved to Svalbard with their family: 'It is much easier here than anywhere else. Longyearbyen is a small community, places are easy to reach and there are so many possibilities for children. My daily life works perfect and is much more satisfying than any of the places I know on the mainland'. When I asked 'And what about your scientific work?', she replied, with determination, 'The opportunities to do Arctic work here are excellent ... We can use the fjords close by as field laboratories and take samples throughout the year, which is an unique option for an Arctic location'. With such an excellent life, I wanted to know whether she planned to stay in Svalbard for the rest of her life. She laughed and responded: 'A couple or more year, perhaps ... It depends very much on my family. There will come a time when my children will get too old to enjoy living here anymore and that may give me a limit'. ∎

OLE MAGNUS RAPP, TROMSØ

The Arctic has been and still is the focus of media attention. There are numerous articles, reportages, websites, and films on the Arctic, which makes it difficult for researchers to keep up with all the information generated. Has all that is necessary to reveal been presented? The answer is obviously no. On board the research vessel the *Helmer Hanssen* I met the journalist Ole Magnus Rapp. It was a bright day and the panorama of the Svalbard coastline was looking superb. We talked on the upper deck. Ole Magnus currently works for the prestigious Norwegian newspaper *Aftenposten*. He is the only employee of the office for northern Norway, which is situated in Tromsø. From there, he keeps himself and Norway posted on what takes place in northern Norway, Svalbard, and north-east Russia. He joined us to report on the science carried out onboard by members of the EU project Arctic Tipping Points. Few, if any, have been to Svalbard more often as a journalist than Rapp: he

estimated that 'Over the last 20 years I may have been in Svalbard 50–70 times'. he added:

I was able to join the Norwegian prime minister and minsters when they visited Svalbard, I joined research cruises and conferences, and I was present after accidents such as a plane crash, a mine fire, and tragic polar bear attacks. Svalbard has changed a lot. There are many more tourists now than 20 years ago and Longyearbyen has become quite commercial. It had got an urban touch.

When I asked: 'Has the interest in Svalbard and the Arctic increased?', he replied 'Indeed, the generic interest in this region is on the increase. There was also considerable interest by the media when the International Polar Year and other large projects started a few years ago, but when the results appeared, the interest declined'. He added: 'The average citizen knows probably little of what research is carried out here, but ought to be interested. It is difficult to explain that something taking place as far away as Svalbard has consequences for one's life. The media have a job to make this clearer'. I agreed and commented that scientists (in Svalbard) have also a job to do in order to make themselves better understood. ∎

JAN MARTIN BERG, LONGYEARBYEN

The Galleri Svalbard is located in Nybyen, the innermost region of Adventdalen in Longyearbyen. It presents artwork inspired by nature, light, colours, and contrasts of Svalbard. Galleri Svalbard is a small gallery compared to those on the mainland but when related to the size of Longyearbyen one can be impressed by the size and quality of both the Galleri Svalbard, and the Svalbard Museum. I met the manager of the gallery, Jan Martin Berg, who invited me into the building. Galleri Svalbard houses the permanent art collection of Kåre Tveter, the Svalbard collection (maps, books, and local currency notes), and a La Recherche collection (lithographs from the 1838 and 1839 expeditions). In addition, the gallery sells art and items of craftwork produced by local artists and houses permanent and temporary exhibitions. Berg informed, 'My job consists of three different areas of responsibility. I manage the gallery and its permanent and temporary exhibitions. I am responsible for the artists and craft designers that rent ateliers and sell their products in this building and I am in charge of the lodgings and ateliers used by those artists that wish to work in Longyearbyen.' I asked him, 'How long have you had this job?', and he responded: 'It is a permanent job and I am employed by the municipality. More and more jobs in Longyearbyen are offered permanently in order to obtain greater stability in the community.' When I asked, 'And what brought you up to such a specific place?', his answer was straightforward: 'Nothing … I applied for a permanent job in the field of art and exhibitions that interested me a lot. I got the job and thus I came here.' I then left the gallery to march all the way to the present-day centre of Longyearbyen, along abandoned mines, a turbid spring-flood river, attentive grazing goose and a cute polar fox that limped through screes – a mix of culture and nature. I hope that the many visitors to Longyearbyen appreciate that up the valley Adventdalen there is an art gallery that is worthwhile visiting. ■

JOHNNA HOLDING, MALLORCA

Doctoral candidates have accomplished much of the recent research carried out on and around Svalbard. From the many that have made the archipelago into scientifically the best-known region in the Arctic, I asked Johnna Holding about her background. She stated: 'I am from outside Washington DC, but I am doing a PhD at the Mediterranean Institute for Advanced Studies, situated in a small town outside of Palma de Mallorca. I work with how organisms are affected by climate change.' I asked her 'How many times have you been in Svalbard?' and she informed, 'I have been there six times, for experiments in Longyearbyen and on-board the R/V the *Helmer Hanssen*. I love to be here, especially on the cruises, because you see a little more of Svalbard. But it is pleasant to be in Longyearbyen too, because you can experience how people live.' Then I asked, 'Can you imagine living up here for a couple of years?' She replied: 'I could foresee living here for a year or so. As I have been here only in the summer it would be nice to see how the seasons change.' When I asked, 'How do you foresee your future?', she responded: 'I would love to continue doing research, particularly in the Arctic. Once you start, it sticks to you.' Thus, science and life is now entering a new phase. Many young researchers study abroad, they have friends from a wide range of countries, they speak several languages, they have often partners who are neither from their original home nor their place of study, and they carry out fieldwork at still other places. This new generation takes a global attitude for granted, but often finds that specific cultural experiences that the older generations worked hard for, despite their global uniformity, are not readily obtainable. Arctic research has become one of the most unifying activities on the globe. What a privilege it is to work there! ■

VIII. SVALBARD LIFE IN ART

ARTISTS AND SCIENTISTS: ANTAGONISTS OR COMPATRIOTS?
FRANÇOIS-AUGUSTE BIARD
BARTHÉLEMY LAUVERGNE
FRANZ WILHELM SCHIERTZ
HANS BEAT WIELAND
WITOLD LOVATELLI-COLOMBO
LOUIS TINAYRE
MARIUS BORREL
FRIDTJOF NANSEN
MICHALOFF WIGDEHL
GERT JYNGE
HEINZ KÖHLER
ANDREI ALEKSEEVICH YAKOVLEV
KÅRE TVETER
TERJE ROALKVAM
VEMUND THOE

Artists and scientists: antagonists or compatriots?*

Human health (in its all-embracing connotation) and societal progress depends upon how ever-present, underlying tensions are treated and dealt with. Disregarding these fundamental strains results in reduced well-being and in impediments. How do scientists and artists, who often are considered antagonists or complete strangers to each other, tackle these basic challenges? Artists have a healing impact upon society by making disturbing issues, and suppressed or zeitgeist matters, conscious through art. Their wider audience often perceive contemporary art as provocative and detached from reality or bizarre. Art exhibits issues and materials that previously have been hidden. Through art, unfamiliar themes can become recognisable and perceptible, and familiar themes are rephrased and subsequently appear in new settings. The shaping of the unfamiliar, the moulding of the unknown – with all the challenges, provocations, and scandals associated with art – supports, by and large, a societal healing process. The cataclysm of conflict-creating topics is diminished and art provides an approach to obtain an instrumental attitude to conflicts, ahead of the time when they become apparent to the general public. A society that does not open up for art and artists does not profit from the healing power of airing preconscious or unconscious issues. Such an artist-induced healing process creates a better, more balanced, and adaption-welcoming society.

Scientists, although applying a distinctly different, seemingly contrary methodology, play a similar role. To provide knowledge about previously unknown processes and answer daunting questions gives scientists a therapeutic role, alongside with other alleviating professions: 'The truth will set you free', according to the apostle John (8:32). This is not only a credo for such distant undertakings such as Christianity and psychotherapy, but also for science and art. It is our endeavours to come closer to the truth that make the world a better place, provided that the central characters can communicate the acquired knowledge in a comprehensible and enthusiastic manner to fellow scientists as well as citizens. Artists and scientists that disseminate their work in an attractive and stimulating manner will stimulate a healing process by raising preconscious and unconscious matters towards public awareness. They cannot do more. The decisive implementation work of the imparted knowledge is the task of the people. Moreover, in a democracy the task belongs to the people's responsibilities: implement knowledge or bear the consequence of the lack thereof.

It is commonly held that art is for art's sake. Further, frequently also science appears to be for science's sake. Scientists can be rightly criticised for this seemingly arrogant attitude. However, politicians who argue that science exclusively has to serve the present-day needs of the people and constrict science through funding straitjackets towards a defined societal goal, can easily make a key fundamental mistake. Applied research is always based upon the science of yesterday. Without a seemingly useless, free science, the foundation of a society is damaged and its future will be in dire straits. Dominance by applied research would soon give raise to declining prosperity, in contrast to well-meant intentions. In addition, dominance by applied research would resemble the well-intentioned ideology of social-realism, which in the absoluteness of former communistic countries left behind cultural desolation. Management research is a necessary tool for a prosperous society, yet giving priority to management research by reducing free research – the key domain of universities – is a secure route to decreasing a country's prosperity. There is a fine balance between applied and free research, and between applied and free art, which has to be considered and argued for against the destructive effect of the seemingly honourable, but harmful, dead-end attitude of 'serve the needs of the people'. The needs of the people deserve better than commonly applied political straitjackets.

The avant-garde, executing seemingly meaningless art or incomprehensible science, sets up the future. This pursuit, often considered disconnected from society or regarded as useless, improves societal subsistence and strengthens man's all-embracing health. Many modern concepts in science are common to literature, art, and psychology: e.g. chaos, resilience, non-linearity, self-organisation, thresholds, and tipping points. Hence, these are not terms for the initiated, but for humanity. Illuminated, defined, and comprehended, they need – in addition to the experience of processing problematic and incomprehensible matters – to be placed into the toolboxes of everyday life. Both scientist and artists apply their laboratories where they invent or create their science or art. Every now and then, a thought lightens the midst of a long night. In the laboratories of life, progress is both simple and simultaneously highly complex. The primitive or primordial is humanity's best friend and it knows how to converse with the soul. Such conversations bring progress to the world.

How, then, have painters in Svalbard contributed to present the unknown? How have they contributed to healing the inevitable wounds of civilisation? In this chapter I present a bouquet of graphics, paintings, and sculptures by fifteen artists that have been inspired by and worked in Svalbard.

Before photography made its entry as the dominant form of documentation, painters were often involved in expeditions. Michaloff Wigdehl (see page 267) travelled to Svalbard several times as an expedition artist. On page 248 he is shown painting the mountain Alkhornet in the outer part of Isfjorden, while onboard the motor ship the *Laila* in 1910, as a member of the expedition led by Gunnar Isachsen. ∎

---

* 'It is only through science and art that civilisation is of value. Some have wondered at the formula: science for its own sake; and yet it is as good as life for its own sake ... Thought is only a gleam in the midst of a long night. But it is this gleam which is everything' (H. Poincaré, 1902).

The French genre painter François-Auguste Biard (1799–1882) was born in Lyon. Biard was the first prominent painter to visit Svalbard, make drawings, and then prepare major paintings after returning home. He travelled around the world, sketching en route. He was particularly successful in rendering burlesque groups, based upon real scenes or people. The paintings often had a 'story' or a setting that appealed to the 'romantic' and compassion-inducing spirit of the time. He portrayed reality with a pinch of salt. Landscapes representing extreme nature became one of his specialities. He is most known for his paintings from the coast of Africa, which illustrate the slave trade. Biard was a dedicated abolitionist.

For a short period, Biard worked as a drawing teacher for the French Navy, but preferred to continue as an independent artist. However, his connection to the Navy ensured that he was invited as an official artist to the La Recherche expedition in 1839. He received many commissions from Emperor Louis-Philippe. At one time he worked in Brazil, where the Brazilian emperor made him a minister for the fine arts. François-Auguste Biard was a true cosmopolite!

In this painting *Magdalena Bay*, which currently hangs in the Louvre, Paris, Biard revealed his talent for dramatic landscapes (page 250). He visited Magdalena Bay (in Spitsbergen) in the summer of 1839, but painted the picture in Paris. The magnificent but queer northern light (he never experienced it) was added in Paris according to his imagination. Probably, he never saw the aurora borealis. The frostbitten landscape, with God's own miraculous fireworks covering the entire sky, was not petrifying enough to make the public shudder sufficiently, so he painted a graveyard in the forefront (reality), with corpses of whalers and hunters lying just beyond (imagination). Clearly, Biard wanted the picture to induce terror-stricken feelings in the spectator. In this respect, he succeeded, because the painting induces shudders and draws attention to the meaninglessness of our short lives and aspirations. However, God's fireworks flicker over the graveyard, forever!

François-Auguste Biard

Biard's sense of the dramatic, which is portrayed through exaggerated scenes – that could be described as burlesque – is vividly illustrated in his paining *Fight with Polar Bears*, which is housed in the Art Museum of Northern Norway, in Tromsø. The painting depicts three mariners in a half submerged boat under simultaneous attack by, which would have been (and is) a rather unlikely event. The sailors seem to be coping well with this incredible situation. One bear has taken hold of the leg of a horrified sailor in a red coat, but a brave youngster is attacking the bear with a simple knife, while the third sailor, with great determination, is on the point of bringing another attacking polar bear's life to an end with a spear. The icebergs surrounding the scene look mysteriously similar to more recent fantasy comics from planets yet to be discovered, while a simple hat rests on the ice floe to the right-hand side of the scene.

In his time, François-Auguste Biard was a most prominent artist in France. His apartment at Place Vendôme 8 in Paris was also a kind of museum and meeting place for artists and explorers. His paintings from the High North afforded him much success in Paris. He was commissioned to paint murals with motives from polar regions for the Museum of Natural History in Paris, situated in the outskirts of Jardin des Plantes. It is rather surprising to find oneself exposed to vivid Svalbard landscapes and reindeer in the centre of Paris, even today! ■

Barthélemy Lauvergne

In concert with François-Auguste Biard, also the French painter Barthélemy Lauvergne (1805-1871) visited Svalbard in 1839. He was born in Toulon and died in Carcés, Provence. Lauvergne is best known for his lithographs and prints. He was also a most experienced man when it came to travelling around the world.

Between 1826 and 1829 Lauvergne served on the ship *L'Astrolabe*, which sailed along the coasts of Australia and New Zealand. Although officially appointed as a secretary to the explorer and naval officer Julius Dumont D'Urville (1790-1842), his artistic talents were used to record zoological specimens and views of New Zealand during the three months that they cruised round the coast in 1827. Later, he spent the years 1830-1832 working as an artist on board the ship *La Favourite*, accompanied by the navigator Cyrill-Pierre Théodore Laplace (1793-1875). The ship anchored in Hobart on Tasmania, where Lauvergne visited Van Diemen's Land, and later on during the voyage he visited Sydney, Australia. Due to his early depictions of landscapes and people, Lauvergne is a highly appreciated artist in New Zealand and Australia today. On the trip to Svalbard with the La Recherche expedition, he was a most productive artist and an entire sequence of graphics derived from his hand can be seen on pages 66, 70, 73 and 179.

Both Lauvergne and Biard were part of a tradition that dominated paintings in Svalbard until the time of World War I, illustrating landscapes, expeditions, and biota. Some of their etchings were later reproduced as lithographs in order for them to be printed in colour. In addition, photographs were taken during the expeditions and subsequently used by the artists when painting in oils or drawing with pastels after they had returned home. Thus, illustrations and artistic unfolding went side by side. ■

252 | Svalbard Life

Franz Wilhelm Schiertz

Franz Wilhelm Schiertz (1813–1887) was a German-Norwegian painter and architect. He was born in Leipzig, but buried at the famous, small village of Balestrand in inner Sognefjord, in Western Norway. In Dresden, Schiertz was a student of one of the most prominent painters of the German Romantic era, the Norwegian Johann Christian Klausson Dahl. Early German Romantics tried to create a new synthesis between art, philosophy, and science, looking at the Middle Ages as a simpler, more integrated period. Their primary interest was contemplation of nature. They aimed at realistic presentations, but attempted to induce an inner emotion in the viewer. They addressed the hidden side of existence, which is considered the reality. By contrast, the exterior was regarded as false, but showed the beauty of the real world that rests within the realm of ideas. The human heart was deeply involved, but not through exaggeration in the same manner as French Romanticists, such as Biard, but by addressing man's pantheistic, religious depths.

Schiertz became a successor of the Romantic German school of painting when the movement was already in decline. In 1836 Schiertz was sent to Norway to make architectural sketches of the ancient stave churches in Borgund, Heddal, and Urnes. In 1837 these drawings were published in a book by Dahl dealing with Norwegian wooden buildings. In 1841 the stave church of Vang was sold and Dahl attempted to keep the ancient building in Norway, albeit in vain. King Friedrich Wilhelm IV of Prussia bought it and Schiertz was commissioned to document the demolition, transport, and re-erection at Bad Warmbrunn in Lower Silesia (present-day Cieplice Śląskie-Zdrój). Schiertz supported the European Revolutions of 1848 (known in some countries as the Spring of Nations, Springtime of the Peoples or the Year of Revolution, a series of political upheavals throughout Europe in 1848). He had to leave Dresden, in concert with many other intellectuals. In 1851 he moved to Bergen in Norway, where he was commissioned by Dahl to make an accurate drawing of the famous former royal residence, Håkonshallen.

In the years 1876 to 1878 Schiertz participated as a landscape painter during the *Den Norske Nordhavs-ekspedisjon*, the first Norwegian scientific expedition that also visited Svalbard. His enchanting picture *Morgen ved Norskeøyne* (Morning at Norskeøyne), depicts an island group in north-western Spitsbergen, located at a latitude of almost 80 °N (page 254). The painting – which currently is displayed at the Art Museum of Northern Norway, Tromsø, but belongs to the National Gallery in Oslo – reflects Schiertz's naturalistic, but also Romantic style that captures the fresh, crisp morning hours of Svalbard's coastal zone (page 254). He continued the tradition of research and expedition illustrations, but his paintings and etchings are more than that: they have also a value in themselves. They are realistic, but also reflect the soul of the painter, which is transferred to the viewer. This results in a mystic dimension and a particular communication that is characteristic of German Romantic painting. *Morgen ved Norskeøyne* reflects the vistas of the expedition, but it is also a work of art in its own right. There is no additional 'story', as in Biard's paintings. The goal is not to make the viewer shudder, but rather to convey the sense of a crisp, clear, and radiant morning atmosphere in Svalbard. It indicates the divine creation, set against the deception of human civilisation.

Schiertz's mysterious watercolour of ice floes is untitled and is on display at the Norwegian Polar Institute, in Tromsø. It provides an ambience of calmness that is both close to reality, but also communicates the enigmatic character of Arctic landscapes. The watercolour shows similarities with some of the works by Barthélemy Lauvergne. ∎

Hans Beat Wieland

Hans Beat Wieland (1867–1945), a Swiss painter best known for his realist images of Alpine scenes, was another artist that visited Svalbard. Wieland left school in 1883, preferring a career in painting to a school degree. In 1885 he moved to Munich, where he studied at the Academy of Fine Arts. Together with Michael Zeno Diemer (1867–1939), he painted a large panoramic scene for the 1893 World Fair in Chicago. In 1894 Wieland joined the Munich Secession, one of many groups that broke away from the realistic tradition, in particular historic genre paintings.

Wieland belonged to a certain type of painters at the end of the 19th century, who with great skill documented historic events, specific scenes, and panoramas. Despite of their artistic qualities, the pictures had a concrete mission that is more or less unknown in today's world of colour photographs, films, and TV, namely documentation through paintings. As a member of this artistic tradition, Wieland reported from Svalbard. During the period 1896–1906 Svalbard was the focus of an entire group of such artists, a few of which are presented in this chapter.

In 1896 Wieland travelled to Svalbard to witness the take-off of Salomon Andrée's balloon expedition to the North Pole (page 257), the hottest media event in those years (pages 182-185 and 210). Wieland had been sent to cover the story by the German newspaper *Illustrirte Zeitung*, and he secured a special deal with the captain of the Norwegian steamer the *Erling Jarl*, which was sailing to Svalbard with tourists on board. Wieland's picture *Danske Bay, Spitzbergen, 24./25. Juli 1896* shows Virgohamna, the Cape Canaveral of the 19th century (page 257). It also shows Andrée's balloon hangar, additional buildings, the supply ship the *Virgo* (from which todays name of the bay is derived), and the deck of the coastal steamer, the *Erling Jarl*. With its vivid colours, the bright atmosphere, and the lovingly painted detail, the painting is by far the best depiction of Andrée's fabled expedition. In the background, we see the mountains and glaciers of Smeerenburgfjorden. Wieland dedicated his painting *Erling Yarl im Packeis, 1896* to Captain Bade, to thank him for the special arrangement that made the trip possible. The picture beautifully depicts the classic ship as it carried tourists to the place at the centre of the world's attention, namely Salomon Andrée's balloon expedition to the North Pole. Are these pictures art, artistic illustrations, or advertisements?

Hans Beat Wieland returned to Virgohamna in 1897, but too late to witness Andrée's balloon lifting off into oblivion. After that, Wieland never visited Svalbard again, but his posthumous archive of papers indicates that he carefully noted the events of 1930, resulting in the recovery of the corpses of the three explorers and the reconstruction of the ill-fated expedition. During World War I Wieland worked as a painter for the Army Museum at Vienna. Glorious war action, but probably not the killing fields, had to be properly depicted. In 1918, he moved back to Switzerland. His vividly coloured images, among them a considerable number of Alpine scenes, went on display in museums, private homes, military garrisons, and railways stations, were acclaimed by critics and became very popular.

Both paintings shown here are privately owned and not on public display. ∎

Witold Lovatelli Colombo

The life of the Italian painter Duke Witold Lovatelli-Colombo is shrouded in mystery. It has been impossible to obtain even basic information about his life. He participated in Prince Albert I's expedition to Svalbard in 1898. It is unknown how Prince Albert and the Duke came to meet, where the Duke was educated as a painter, and what he painted before or after the expedition. His duty on the cruise was to make illustrations of marine invertebrates, because they quickly lose their colour after death, when exposed to air and increasing temperatures. Duke Lovatelli-Colombo made also landscape drawings and oil paintings in Svalbard.

Shown here is *Mont Temple, Baie Sassen, Spitzberg 1898*, an impressive aquarelle in different shades of ochre and white. The rocky upper part is nicely reflected by the shivering fjord surface and the picture opens up to the left towards the wide Isfjorden. The mountain range Tempelfjellet is clearly impressive. Also shown is the aquarelle titled *Glacier Post, Sassen Bay, Spitzberg 1898* (page 258). It is one of many depictions of glacial fronts that always impresses even the most experienced visitor or researcher. A tower of bulging ice, mirrored in the surface of the adjacent fjord, creates respect. Both paintings belong to the Musée océanographique de Monaco, but are not on display. ∎

Louis Tinayre

Louis Tinayre (1861–1942) was the son of French Communards and spent eight years of his childhood in Hungary after his parents had fled France. He studied painting at the Budapest Academy, and later returned to France in 1880, where he soon established a reputation as an artist, specializing in newspaper reportages. In 1895 he was sent by Le Monde Illustré to cover the quelling of an uprising in Madagascar, then a French protectorate, by French troops. The country fascinated Tinayre. He spent six months there alone, and produced many drawings, photographs, and illustrations. On his return to France, these works became the basis of a set of eight huge paintings (3.5 x 5 m). In 1898 he returned to Madagascar to work on a giant panorama of the capture of the town of Antananarivo by the French in 1895. He took with him one of the early cameras, a Lumière Cinématographe, which he used to record the events, in addition to the images that he captured in his sketches. The negatives depict Madagascan people dealing with their everyday lives and the composition of the images suggests that Tinayre used the photographs as aides when constructing his panorama. Tinayre exhibited the panorama at the Paris Exposition in 1900. He then became official painter to Prince Albert I of Monaco, who was not only a sovereign, but also an oceanographer and enthusiast of the use of cinematography on his scientific expeditions, which included Svalbard.

Tinayre is by far the most prolific painter among the many that were commissioned by Prince Albert. He accumulated a wealth of oil paintings and pastels during the expeditions of 1906 and 1907. Furthermore, he often based his paintings on photographic plates, which later were transformed into pieces of art. His paintings are thus copies of reality, but reality was interpreted by him according to his feelings, memories, and psychological projections. More than 35 paintings are known, most of which are on display or in the depot of the Musée océanographique de Monaco and at the Norwegian Polar Institute, in Tromsø. Tinayre also decorated the Grand Amphithéâtre at the Institut océanographique in Paris (an extensive work that reflects Prince Albert's work in oceanography. During World War I, the painter became a journalist for l'Illustration. During the remaining part of his life, Tinayre travelled to Brazil, Morocco, and southern France, where he died in his family hometown, Issoire, in Auvergne.

One evening after finishing work, Prince Albert inspected the most recent painting, which had been made close to a glacier, and said: 'This, my dear Tinayre, is a very good painting. The glacier and bay are outstanding! I recognise the very details'. After a moment of reflection, he added, 'Do you know how I will suggest that this region is named? The Bay and Glacier Louis Tinayre', and this is exactly what happened. Tinayrebukten (The Bay Louis Tinayre) is a small fjord off Krossfjorden, a side fjord of Kongsfjorden. La Baie et le Glacier Louis Tinayre (page 261) can be seen at the Musée océanographique de Monaco.

Also shown is the depiction of cartographic work in Svalbard that Prince Albert I supported. Two researchers are depicted on a mountaintop, taking triangular measurements for the construction of exact maps. It was this work that created some of the preconditions for granting Norway sovereignty over Svalbard in 1925. The motif has been shown on stamps in Norway, but the original can be seen at the Norwegian Polar Institute, in Tromsø. ∎

Marius Borrel

Little is known about the Svalbard painter, François-Marie Borrel (1866–1937), known as Marius Borrel. He was an engraver and painter, but we neither know where he was educated nor what he did after he finished his lengthy engagement as expedition painter for Prince Albert I of Monaco. Borrel accompanied the prince on many cruises in the years between 1888 and 1903, primarily to draw and paint living marine animals. Probably his best-known painting is *La Construction du Sacré-Coeur de Montmartre*, which is currently exhibited at Musée Carnavalet in Paris. At least three of his paintings depict scientists, personnel, and glacier fronts in Svalbard. Borrel's work is interesting because he brought landscape painting to a limit. Despite their artistic qualities, his pictures had the concrete mission to document, in colour, the appearance of the Arctic archipelago, Svalbard. However, to the best of my knowledge, Borrel did not visit Svalbard or any other Arctic region! The two paintings depicted here must be based upon black and white photographs and were probably painted in 1898 (page 262). What do calving glaciers look like in the full colour of reality? *Front du glacier de la Baie Ginevra* (left) and *Rue de glace au Glacier Post* (right) marked the end of the genre of documentary landscape paintings of Svalbard. These two appealing pieces of art are effectively colourised photographs, reproduced as oil paintings. They signify a former reality in which the painter was disconnected from personal experience, the natural ambience, and the upwelling emotions of the real sight. They are virtual creations of an outer reality where the artist could not even project the colours onto canvas from memory. Instead, they imitate an assumed reality. Thereafter, only colour photographs could replace the realm hitherto occupied by paintings.

Both of the above-mentioned paintings belong to the Musée océanographique de Monaco, but are not on display. ∎

Fridtjof Nansen

Among the many facets of Fridtjof Nansen (1861-1930) is that he was a gifted artist that made an impressive number of illustrations and graphics. Surrounded by an environment of art, Nansen had made drawings since he was a young boy and when he was in his early twenties he had considered becoming an artist. However, he finally settled on studying biology and natural sciences. Nevertheless, he continued to sketch and illustrate his own books. He was regarded as considerably talented, and in addition to his many professions he could have added that of a full-time illustrator. He based his works of art on his expeditions, but he made also portraits and an entire series of hunting motives.

Among Nansen's teachers were the painters Franz Wilhelm Schiertz (page 255) and Erik Werenskiold, the latter best known for his meticulous and epoch-making illustrations of the *Heimskringla*, a history of the Norwegian kings by the Icelandic historian, poet, and politician Snorre Sturluson (1179-1241). In Werenskiolds book illustrations Nansen's portrait emerges frequently to exemplify Norse nobility. His friends in the 'Lysaker circle', foremost among which was Werenskiold, influenced his art. The circle included Nansen's brother-in-law, the world famous marine biologist George Ossian Sars, the writer Hans E. Kinck, the folklorist Moltke Moe, the painter Eilif Peterssen, and the school textbook editor Nordahl Rolfsen. Together, they formed a liberal cultural intelligentsia with stark national overtones, which was considered a counterweight against two fronts: the common liberals and countryside nationalists.

Among the plethora of worthy graphics from Nansen's hand (see also page 26), we show here an unpublished one of a naked woman jumping over some Svalbard mountains (page 265). It is a pen and Indian ink drawing, signed Fridtjof Nansen and was given as a Christmas present to Bjørn and Annemi Helland-Hansen in 1929 (Nansen was then 68 years old). Is this young woman a Fata Morgana during long and frustrating trips to the Arctic? Is the subject reminiscent of his younger days? Is it the liberated virgin Norway that claims sovereignty over Svalbard in a jolly leap over a mountain range? Also shown is an unpublished aquarelle (upon a photograph) signed 'FN', which may relate to Fridtjof Nansen's first trip to the Arctic, when he travelled with a sealer to western Greenland in summer 1882. At the time, Nansen was 21 years old. We see Nansen to the left and Captain Axel Krefting to the right, flaying a polar bear on the ice. ■

Michaloff Wigdehl

Jon Michael Andreassen Wigdehl (1856–1921) was born close to Gildeskål, south of the city of Bodø, in northern Norway. His grandfather was the legendary Russian John Russ who arrived in Bodø by chance around the year 1820. Jon later adopted the name Michaloff, underlining his roots and intricate lineage. He received a military education in Trondheim and Stockholm, and while he was in Stockholm he took lessons at a drawing school. During those years he spent much time painting and held his first exhibition in Tromsø, Norway, in 1895. He later went to Oslo, where he attended both the Royal Drawing School and the private school of the renowned painter Harriet Backers. Several exhibitions followed. Wigdehl often spent summer in northern Norway and winter in Christiania (Oslo). He was a deeply religious man, who started each of his paintings by writing in his diary 'Dear Jesus, help me to produce good art'. He saw God's awareness of beauty behind the creation and then attempted to translate it onto canvas.

Probably motivated by particular achievements of Tinayre, the expedition leader Gunnar Isachsen asked the famous Norwegian painter Christian Krogh to recommend a painter that could accompany him to Svalbard in 1910. Krogh suggested Wigdehl, but Isachsen hesitated because of Wigdehl's age. However, following a meeting in person, Wigdehl was invited and accepted immediately. The expedition reached various Svalbard destinations up to a latitude of about 80°N (see Wigdehl painting in open air page 249). Already in 1911 Wigdehl presented his first Svalbard exhibition in Oslo, with 30 paintings. Obviously unaware of Biard, Wieland and Tinayre, to mention some of the earlier Svalbard painting protagonists, Norwegians assumed that Wigdehl's paintings were the first ever created in Svalbard. Wigdehl became renowned as the 'Painter of Spitsbergen'. He returned to Svalbard in 1915, but the trip ended in tragedy when the vessel was shipwrecked in the ice. Fortunately, the crew survived and returned to the mainland.

In 1919, Wigdehl embarked on his last trip to Svalbard with youthful eagerness, despite being 62 years old. On that occasion, the venture was under the leadership of the ever-present geologist Adolf Hoel. In thick fog, they ran onto a sunken rock, but survived. Despite the foggy weather and nearly being shipwrecked, Wigdehl succeeded in producing 13 paintings. In 1920 he held his final Svalbard exhibition in Oslo. With an exuberant will to live, he planned to spend a winter in Svalbard in order to create more canvasses and cover all seasons. Fate did not provide him with this opportunity, as he died in 1921.

Wigdehl always kept to his colourful, warm, somewhat naive, but confronting style that never glanced at the plethora of 'isms' of his contemporaries. He spent time in Paris and had exhibitions in many European cities, but always kept to what seemed right for him. Impacted and deeply fond of the grand nature of Svalbard, the worship of nature was his credo and oeuvre: boundless, without presumptions, unapproachable, and spellbinding.

Both paintings of mountain cliffs and calving glaciers in Svalbard are untitled and are on display at the Norwegian Polar Institute, in Tromsø. His brushstrokes render smooth surfaces that are textured and provide a mineral appearance of colourful earths. Wigdehl portrayed radiant colours, but no flamboyance or artistry. There is not much depth to his paintings; rather, they have a two-dimensional structure. ∎

Gert Jynge

Gert Jynge (1904–1994) was a Norwegian painter and graphic artist. He started his artistic career by attending classes at a drawing school in 1923, with Olaf Willums as his teacher. He continued his education at the Art Academy in Oslo in the period 1929–1931, where he was taught by Axel Revold. In his early days, Jynge presented woodcuts at various exhibitions. His main productive period ranged from about 1925 until about 1948, when he ceased to exhibit his art. However, throughout his life he continued to draw. From the 1950s onwards, he was a highly appreciated drawing teacher at the State School of Art and Crafts (Statens håndverks- og kunstindustriskole) in Oslo.

Jynge belonged to a group of artists whose works were against the mainstream, French-orientated painters of the inter-war period in Norway. In common with many artists in those years, he had close contacts with socialistic-orientated circles. He published woodcuts in leftwing newspapers. In the early 1930s, Jynge was clearly expressionistic in style and this did not fit well with current styles in Norway. Unlike Impressionists, who sought merely to imitate nature to the very last visual sensation and flickering of light, German expressionist painters typically distorted colour, scale, and space to convey their subjective feelings about what they saw. However, World War I had scarred many of these artists for good. As a result, from 1915 onwards, German expressionism became a bitter protest movement as well as a style of art. The centre of much of this avant-garde art was the Sturm Gallery in Berlin. The movement found supporters in many countries that wished to protest against societies that were at odds with their time and with the needs of the majority, poor people. Jynges art was influenced and attracted by German post World War I expressionism.

Gert Jynge's father was a director of the Norwegian railway and had contacts with the Store Norske Kullkompani, the state-owned coalmining company in Svalbard. This provided him with the possibility to visit Svalbard with the first coal-carrying ship to leave northern Norway in spring 1930. Jynge lived in Longyearbyen all summer, and returned to the mainland in autumn. He had particular focus upon the special working environment of the miners in Longyearbyen and their life and work became the sole focus of his art. Much of Jynge's art dating from 1930 and later concerns workers. He seems to have been the only artist that was not dedicated to the overwhelming Arctic nature beyond the archipelago, yet what could be experienced inside Svalbard 'rocked'. There was an element of introspection in the art of that time – What was it like to be a miner? To the best of my knowledge, Jynge's art is probably the only serious Norwegian attempt to provide an artistic expression of mining in Svalbard that reached a public gallery.

In 2010, one of Jynge's 'mining' paintings was bought by the North Norwegian Art Museum and is currently on display in Tromsø: *Kullgruve, Svalbard* (Coalmine, Svalbard) (c.1932; page 269). I also show an aquarelle showing the head of the mine, with miners extracting coal. There is a dismal ambience in both of these pictures. They reflect a gloomy reality and seem to induce the sentiment of difficult working conditions and a hard and unhappy life. Does the dark atmosphere reflect suppressed anger, resignation, or gnashing teeth? One seems to detect a subdued anger of injustice and the obstructed energy of social insult. The oil painting reflects a trench atmosphere, just before the battle against the restraints of a class society. Such a battle did not happen in Norway. A social democratic party came to power in 1935 and introduced a lasting period, in which social democratic ideals, such as the fostering of a progressive evolution to change a capitalist into a socialist economy, became prominent in Norway. Social democracy argues that all citizens should be legally entitled to certain social rights (universal access to education, health care, compensation for workers, childcare and care for the elderly, and trade union rights) and the success of the movement dissolved the atmosphere depicted in Jynge's paintings.

The drawing (page 87) and the aquarelle (this page) are stored at Norsk Bergverksmuseum, Kongsberg, Norway. ∎

Heinz Köhler

There has been a suite of gifted amateur painters, who, stirred by the mighty landscape of Svalbard, were inspired to apply the very best of their creative powers in the production of art. Among these was the German Heinz Köhler (1913–1943). Köhler was an inspector of one of the many secret weather stations that the German armed forces dispatched to Svalbard during World War II. As part of these secret operations, personnel were transported to Svalbard by submarine. In order to fit into a submarine, provisions, winter cloths, tools, photographic and scientific equipment, sledges, skis, communication equipment, weapons, radio probes, tents, building material, fuel, and other items all had to be packed into boxes that were only 55 cm in diameter. The logistics of such operations were remarkable. On 7 October 1942, Köhler left Tromsø on the U377 to work at the weather station named Nußbaum. The six member crew built the station close to Signehamna, a site where it would be invisible from the fjord Lilliehöökfjorden (Krossfjorden). The weather station sent regular weather reports from 30 November 1942 and onwards. Köhler drew weather maps.

As a central European, Köhler was overwhelmed by the might of the landscape, the variability in light, and the natural wildness in Svalbard – all of which is reflected in his paintings. The sunrise after four months of darkness stimulated him to paint the event (page 270). However, even after the sun starts to rise above the horizon in Svalbard the landscape remains wintery for some time. Köhler died too early to experience 'real summer' in the Signehamna region. One of his paintings is a sketch with the inscription '20 h am 17. Mai, Mitra' (Kapp Mitra is a prominent peak in north-west Spitsbergen). The date 17 May is Norwegian national day and visitors that have experienced this expression of national pride and self-confidence know that this day is very special to Norwegians. What prompted Köhler to make the sketch on that particular date? Was it by chance, or does the image inform viewers about Köhler's respect for the intense Norwegian notion of independence and self-determination? We will never know the answer.

The Nußbaum station was detected by Norwegian troops stationed in Barentsburg, and Köhler was shot during an attack on 20 June 1943. The circumstances around his death are not well known, but most of his colleagues were able to escape to Magdalenefjorden and were rescued by the submarine U302. The station was blown up, but surprisingly Köhler's paintings were taken care of before doing so. The Norwegian Defence Museum stores them in the depot at the Svalbard Museum, in Longyearbyen. ■

Andrei Alekseevich Yakovlev

Andrei Alekseevich Yakovlev (1934) is a well-known artist from St. Petersburg. He was responsible for three sets of paintings made in period 1960–1980, on life in the North, entitled 'Spitsbergen', 'Chukotka' and 'Taimyr'. Yakovlev graduated from the Surikov Moscow Institute of Art, Sculpture and Architecture (1956–1960), where he had studied under the supervision of E. Moiseenko. In 1972, he was awarded a silver medal by representatives of the All-Russian Exhibition Centre (formerly VDNKh USSR). He is an Honoured Artist of the USSR and Laureate of the Repin State Prize. An exhibition to mark Yakovlev's 70th birthday was held in the State Russian Museum, St. Petersburg in 2004.

Yakovlev worked in Svalbard in the years 1960–1961 and 1986. Here I show two paintings that Yakovlev worked on in Svalbard in 1986 and which currently are displayed at the Barentsburg museum. Both paintings suggest significant elements of social realism: the focus is not on nature, as such, but the template that human settlements, industry, and modern technology impose upon nature – modernity and culture in the wilderness. Applying technology provides the possibility for human existence in an inhospitable nature. *Mine Pyramiden* was painted in late summer, when the settlement Pyramiden was still inhabited and fully functioning (page 273). The impressive mountain in the background clarifies for the viewer why the settlement is called Pyramiden. The settlement seems almost deserted and only a few figures can be detected. A red star and characteristic Russian catchphrase '*Miri mir*' (Peace to the whole world) with the indispensible hammer and sickle, convey that the settlement is socialistic and belongs to the Soviet Union. The second painting is entitled *Barentsburg*. However, it is not the stretched-out, anonymous-looking settlement or the ungenerous nature that catches our attention, but rather the freighter in the harbour and the helicopter, obviously on its way to the then newly built and controversial Soviet helicopter base at Kap Heen. The painting underlines how man embarks upon the wilderness to make a living supported by technology. It depicts an old theme that was of particular significance in communistic states: the enslavement of nature to serve the needs of humans. Nature was expected to serve, and brute force and heavy technology was needed: '*Communism is Soviet power plus the electrification of the whole country*' (V.I. Lenin). The total neglect of nature is the road to self-enslavement. Where has the battleground against nature really provided a clear victory? Nature was fully defeated in Pyramiden by locating a model city into the wilderness and there was a rather doubtful compromise between nature and human needs in Barentsburg. In support of a sustainable development humankind needs to find a harmonic position inside the niches that nature provides and has to select a policy that takes care of both the individual and the collective.

Kåre Tveter

Kåre Tveter (1922–2012) spent three years studying at Bjarne Engebret's school for painters. He needed time to develop his talents, and at the school he was known as the 'eternal amateur and Sunday painter', but ultimately he became the only professional and successful artist among all of the students. From 1954 to 1957, he continued his education at Statens Kunstakademi, Oslo, under Per Krogh and Alexander Schultz. Tveter received several scholarships in the period 1955–1966. He illustrated books and held separate exhibitions in Norway, and contributed to collective exhibitions in foreign countries. Tveter is known for his paintings of landscapes, especially those with autumn and winter ambiances. He was a master of darkness and subdued shades, the minimal use of colour, and the light of high-latitude winters and the Arctic.

Tveter had particularly strong connections with Svalbard. He first visited Svalbard in 1982 and thereafter he was spellbound by the light and grand vistas of the place. He is often referred to as a painter of light, but, as I will show, he was also a painter of darkness. He painted daringly coloured, light-flooded, and darker than dark images. Direct dialogue between these extremes never turned up in his art. Tveter kept the discourse of the extremes for himself. Alternatively, Tveter may have intended to induce counter-imagery and antitheses in the viewer: a dialogue may spring forth in the viewer's soul when studying Tveter's colourful and dark pictures.

Tveter travelled frequently to Svalbard, but did not paint, sketch, or take photographs there. Instead, he mentally sampled the landscape and absorbed impressions that later resurfaced on canvas and paper in his atelier. His art is represented in many museums, but the most significant collection of his Svalbard paintings can be seen at the Galleri Svalbard in Nybyen, Longyearbyen. It is truly a worthwhile experience to examine his large-format Svalbard landscapes on display there. The entire range of imaginable hues of pastel tones of light, sublime shades, and immense dark are spellbinding. As mentioned earlier, if one is interested in colour, one should head north. In particular, the light during the time of year when the sun returns is mesmerising; at times, it is psychedelic. The colouring of the snow-covered mountains and the horizon of the frozen sea are beyond description for most people. However, Kåre Tveter was able to perceive the vistas, differentiate between the sublime touches, and preserve them in his art. He knew the language and had the expressivity to let the vistas materialise again. He took hostage of these vistas and shared them by creating large canvases and mega-large aquarelles. *Arktisk* månenatt (Moonlit night in the Arctic) – held in the National Gallery in Oslo – is a rather dark, contrast-rich, vertically orientated manifestation of nocturnal mountain flanks and summits (page 274). Viewers can admire the painter's statement that 'light is most insubstantial' – the darkest possible matter joined to the most insubstantial (pages 276–277). The north is associated with incompatible qualities: light (pure and innocent) and dark (dangerous and terrifying). Tveter's art seems to support these conceptions, but he deals with the extremes, not by presenting an incompatible contrast – a dichotomy – but by presenting a dialogue.

Kåre Tveter, the painter of light, was strongly connected to the dark and the twilight: 'Without darkness light has no meaning', he stated. God created light after the primordial earth was formed, 'And the light shineth in darkness; and the darkness comprehended it not' (John 1:5). I suppose Tveter disagreed, because the darkness in his works of art fully comprehends the light. Darkness and contrast did, indeed, represent one extreme in Tveter's universe. However, his paintings can also be lyrical and remind the viewer of almost non-existent, dreamed skylines, where the 'colours are the doings of light, doings and sufferings' (J.W.v. Goethe). Their etheric nature is also revealed by the sensitivity and caution of Tveter's brushstrokes. In his twilight paintings, the vehement Svalbard landscape transforms into a country that appears almost not to exist. The brutality and harshness of the Svalbard scenery can at times resemble a Fata Morgana, a dreamland, a Shangri-La, or mystic Shambhala. Tveter discovered a mystery, analogous to how graceless and clumsy caterpillars are transfigured into airy, light-coloured butterflies. He was able to see what was beyond the outer world. With these particularly sensitive psychological antennas, he was able to sense the compensatory reflections of a brutal, material world. Tveter was truly 'Appolinic'. He discerned the god of light, music, and truth during his visits to Ultima Thule. In Tveter's art, Apollo forgets about the ice-tormented and stony, material world and materialises himself in pastel tones and dim moonshine.

There is no Svalbard life without Kåre Tveter: he was not only the painter of light but also the painter of life that comes into being as 'the inner light' – the mind's natural landscape, the soul's archetypical terrain, and the subtle world of our psyche that so easily is crushed by overexposure caused by contemporary overloads of visual stimuli. Like the trolls, his art shuns the glaring light of intellectual consciousness and expresses itself in the transparent, subdued, and wary – inner landscapes, where white is not a colour, but the name of an entire range of shades. ■

'In the colourful reflection we grasp life', (J.W.v. Goethe)

Terje Roalkvam

Terje Roalkvam is a contemporary artist mostly known for his sculptures. He was educated at the State School of Art and Crafts (Statens håndverks- og kunstindustriskole), Oslo (1967-1971), and the Akademii Sztuk Pieknych, Warszawa (1971-1972). He has had numerous solo exhibitions and has participated in many joint exhibitions in Norway and abroad. He has participated in research cruises around Svalbard and the Barents Sea, and for a short period he lived in Ny-Ålesund. His Arctic experience has influenced his recent work, resulting in the exhibition Vippepunkter (Tipping Points) in summer 2011, at the Galleri Svalbard, Longyearbyen.

Throughout this book, with its particular perspective on life in Svalbard, it should be noted that over time humanity has become exposed to an increasing contrast between nature and technology and/or culture. This contrast sustains a balance–imbalance cycle. In turn, the cycle appears to be at the base of Roalkvam's sculptures. His works weigh the contrasts light and dark, rough and smooth, warm and cold, and nature and culture. Thus, the basic forms of the sculptures are divided into two parts that are never symmetrical, but a concord between two opposites that respond to each other. Life is never perfect and, as in Japanese ceramic objects, there is always a small deviation from perfection (in order not to irritate the gods). Stone or wood represents nature in Roalkvam's work, whereas culture often is embodied in metal or a material of industrial significance (such as aluminium and coal). Both parts are forced to coexist with one another, but they also move away from each other. A coalition with permanent conflicts and friction, similar to man who depends on Nature but for the most part lives in an antagonistic relationship with her.

Although Roalkvam's sculptures are static and fixed forever, the resulting processes into which the viewer is entangled is a dynamic one. There is a pendulum in the process that the viewer is invited to engage in, balancing around a threshold of, for example, nature-culture, natural-industrial, even-uneven, and light-dark. Roalkvam's sculptures are gentle. He does not intervene in the viewer's soul to transfer his message, for example, through vehement video instalments. He leaves the viewer in a state of freedom regarding whether to enter the process. He invites us to ponder about the antagonistic directions and processes, i.e. the thresholds of man's existence. Away from the struggles of daily life, humanity is concerned about this threshold: how can we balance nature and culture? Where are the tipping points that either knock us out of balance or keep us in balance? How can we strengthen our resilience in order to keep within the range of states in which sustainability is possible?

Shown here are three sculptures that combine aluminium with red Svalbard rock ("DIVISO III-XI" 2011; page 279), brown Svalbard and aluminium ("DIVISO IV-XI" 2011, left) and coal and aluminium ("DIVISO V-XI" 2011, right). Two worlds fused into a unit and an underlying conflict that can only become productive through a dialogue – divisional, but restrained into one. The dynamics of systems function like that: oneness that is divided into parts. ∎

Scientists of the Arctic marine ecosystem research network ARCTOS, which builds bridges between senior and young scientists, national and international research, science and industry, science and culture, regularly invite young artists to participate in research cruises to Arctic waters around Svalbard. In cooperation with the Tromsø kunstforening ARCTOS also provides the artists with the possibility to display their art through the annual POLART exhibition. It has long been known that the roots of science and art are not entirely separate. Members of the ARCTOS community strongly believe that artists and scientists are compatriots that have to learn how they, either individually or together, can shape a healthier society by moving the unknown to the sphere of the conscious. 'Incubated' on a ship, the traditionally separated brethren experience a setting where joint understanding and mutual fertilisation can and should take place. The old tradition of having artists involved in research cruises is thus continued. Today, artists are no longer the illustrators of landscapes, sights, and organisms, but in one or other way are connected to or interested in scientific and artistic processes.

Among the more recent artists to be given the opportunity to join a research cruise to Svalbard is Vemund Thoe, who, in terms of style, identifies himself as a neo-expressionist. Neo-expressionism is a style of modern painting and sculpture that emerged in the late 1970s. It developed as a reaction against conceptual and minimal art. Neo-expressionists are sometimes called 'New Fauves'. Fauvism is the characteristic style of les Fauves (French for 'the wild beasts'), a short-lived and loose group of early 20th-century modern artists. The 'new wild ones' returned to portraying recognisable objects, often in a rough and violently emotional way and applying vivid colours. Thoe was educated in Oslo, Trondheim, and London. He participated in a wide range of group exhibitions and several solo exhibitions. Thoe has remarked that neo-expressionism affords him a sizeable span of possibilities and techniques with which to articulate his visions, intuition, and creativity.

Shown here is an imaginative visualisation of the research vessel Fram in August 1896, just after it was re-

Vemund Thoe

leased from a three-year long, close to deadly embrace by the Arctic Ocean ice. Released from the mortal grip, the Fram is shown heading south towards the closest settlement, Virgohamna, while Andrée is preparing for the North Pole, which the Fram failed to reach. Thus, the image depicts the most advanced and successful Arctic expedition of that time meeting its even more advanced successor, which will soon be doomed to disaster. The ship, battered by ice, pressure ridges, and ice storms, is anchored in front of the glacier Frambreen, opposite Virgohamna and Smeerenburg. The size of the glacier front seems to intimidate the brave vessel. The danger that was escaped only a few days ago seems to linger around the ship, as reflected by the gloomy nature of the applied colours. We can sense an expression of how the ship and its crew felt after surviving the sheer impossible.

The main painting, based upon the artist's photograph of the front of Kongsbreen (page 280), is brighter in nature compared to the portrayal of the Fram. It designates the airy atmosphere of a day with bright sunshine through a partly cloud-covered sky. The cracks both in glacier front and in the bluish glacier portal are reflected in the calm waters in the foreground. Can one hear the treacherous silence of the glacier front? One awaits the rolling thunder of an ice cascade – light and pure, dangerous, and horrifying. Seemingly silent and static, every glacier is a creeping and thundering titan. Thoe exposes us to two ambiences, the present silence of our short life and the thundering vehemence of geological timescales. The paintings are owned by the artist. ■

IX. CLIMATE CHANGE IN THE ARCTIC: WHAT IS AHEAD OF US?

WARMER AND WETTER
AMPLIFICATION OF GLOBAL CLIMATE CHANGE
SENSITIVITY TO ENVIRONMENTAL CONTAMINANTS AND RADIATION
SEA-ICE DECLINE IS A THREAT TO SPECIES
A WARMER OCEAN CHANGES ECOSYSTEMS
THE INESCAPABLE ACIDIFICATION OF THE OCEAN
LANDSCAPES AND ANIMALS
A PASSION FOR PRISTINE ENVIRONMENTS: ARCTIC TOURISM
THE INFRASTRUCTURE CHALLENGES
ADAPTATION TO CHANGE

It is widely known that the Arctic changes rapidly and faster than anywhere else on the planet. In 2007 the sudden decrease in summer ice resulted in a public outcry in the media. A new record low sea-ice cover was recorded in 2012, but that time the news hardly made the headlines. There was so little ice in August and September that ships could have passed straight across the Arctic Ocean to and from the Far East. A new era of ocean transport seems to be appearing on the horizon, one when larger quantities of cargo, oil, and gas will be using the short connection across the Arctic Ocean between Europe and north-eastern North America en route to the north-eastern Asian markets.

In response to the dramatic increase in the effects of climate change, in recent years there have been several major evaluations of climate change in the Arctic region. In this chapter, elements of some recent studies are presented, with emphasis on Svalbard.

Warmer and wetter

In order to determine how the climate will change in decades to come and what the consequences may be for people's livelihoods, the United Nations Intergovernmental Panel on Climate Change (IPCC) has developed various future climate scenarios. These reflect climate gas emission scenarios that will depend on the development of the human population and the manner in which essential resources are used. Today's political trends seem to favour a scenario where economic growth appears more essential than environmental protection and social equality (scenario A). This scenario predicts a population growth of up to 15 billion by the end of the 21st century. Technological changes that would alleviate the challenging consequences of climate change are assumed to take place slowly. It is estimated that by 2100 only 28% of energy production will come from sources that do not emit carbon dioxide. Further, the average global temperature of the atmosphere is expected to increase by up to 4 degrees Celsius.

An alternative scenario (B) assumes that the world is more favourable to environmental protection. According to this scenario, the world's population will increase to approximately 10 billion by the year 2100 and the level of economic development will be moderate. Coal will contribute 22% of energy production, and 49% of the production will derive from sources that do not emit carbon dioxide. The average global temperature is expected to increase by up to 2 degrees Celsius. This scenario becomes less and less likely.

Both scenarios predict that Svalbard and adjacent regions will face increases in annual average temperatures in the range of 3–8°C towards the end of the century, **which is expected to be well above the global average.** All seasons will experience temperature increases, which will be greatest during autumn and winter. Also, they will be greater over land than over the ocean. Precipitation will increase in all seasons, but mostly in autumn and winter. However, regional differences will be significant. Towards the end of the century, the season of snow cover will be greatly reduced, and, as already detectable, extreme weather events, with strong winds and intense precipitation will increase in frequency. There will be gradual warming of permafrost, with a possibility that the melting may become irregular due to extreme atmospheric temperatures.

In essence, the climate will become warmer and wetter in Svalbard in the future. ■

Amplification of global climate change

Self-amplifying mechanisms in the Arctic increase the effects of global climate change. Ice and snow have light-coloured surfaces that reflect sunlight and thus counter some of the effects of warming. Increases in temperature result in ice melt and shorter periods of snow cover, and this will also contribute to increased warming, as the surface will darken. Increasing amounts of soot are deposited in the Arctic, which derive primarily from coal burning and forest fires. Such soot deposits will further intensify ice and snow melt. However, soot sources can be reduced, either by application of technology or by changes in agricultural and forestry practices. The temperature of the Arctic Ocean is also influenced by the Gulf Stream, which transports copious volumes of warm water into the ocean. It is a delusion to imagine the Arctic as far away and on the periphery. The Arctic Ocean is closely connected to the entire northern hemisphere and thus cannot be considered in isolation.

Clouds are an important part of the climate system, but their role in the Arctic is complicated, not well understood, and consequently not appropriately represented in models. Good predictions of the future cloud cover in the Arctic are thus difficult. In addition, calculations of

the effect of climate change upon currents in the ocean and atmosphere are not reliable and more knowledge is needed to understand better how these key processes in the global climate system will change.

A major source of amplification in the Arctic is the emission of methane, which has a strong impact on climate. Significant amounts of methane are stored in a frozen state (as gas hydrates) in Arctic marine bottoms and wetlands that often have permafrost immediately below the surface. Warming seawater or melting ice on land may thaw permafrost and consequently released methane can be emitted to the atmosphere. The total amount of amplified emission is not well known at present, but calculations indicate that methane may be the most significant source of future global warming.

In essence, although we recognise the trends in the development of the climate in the Arctic, lack of adequate knowledge regarding self-amplification renders any projections less reliable. Warming may develop quicker or slower than forecasted at present, but experience suggests that the development is more rapid than projected. ∎

Sensitivity to environmental contaminants and radiation

Climatic change makes the Arctic more vulnerable to contaminants and ultraviolet radiation, due to the general northwards transport of pollutants from the highly-populated low latitudes to the less-populated high latitudes. In turn, the otherwise pristine Arctic, with only a few local contaminant sources, is transformed into an undoubtedly polluted region. The pollutants are either deposited or accumulate in the atmosphere towards the north, as reflected by the ozone degrading substances that are the cause of the hole in the ozone layer above the Arctic. However, also currents, such as the North Atlantic and Norwegian Coastal Currents, supply the Arctic with a suite of contaminants from the industrialised areas to the south.

As a result of various conventions, the production, consumption, and emission of many contaminants have strongly declined internationally. There has since been a decrease in the levels of many of the main contaminants measured in several species of Arctic mammals, although concentrations are still high. Some of the contaminants analysed in recent years are still showing signs of increase in some Arctic regions, but are decreasing in others. Further, new contaminants are continuously produced, for which appropriate analytical methods have to be found in order to identify their impact. These are probably not as harmful as the earlier ones, but there is much insecurity due to limited knowledge and analytical capacity.

Changes in the climate system may impact contaminant transport to the Arctic. The transport of such contaminants could be intensified, dependent on how the involved mechanisms alter. Meanwhile, warming may not only cause the Arctic to function as a sink, but also a source of contaminants known as persistent organic pollutants may become exposed to revolatilisation by climate change. Some contaminants that previously were included in permafrost, glaciers, and sea ice could be released and increase the contaminant concentrations in meltwater. Climate change that may be responsible for forest fires and the release of gases and soot into the atmosphere could be yet another source of contamination of the Arctic. Also, we should not overlook the unidentified stress that a cocktail of various contaminants has on organisms. Arctic organisms that are already stressed as a consequence of climate change (e.g. higher temperatures, lack of ice, and lack of adequate food) become more exposed to additional stressors, such as contaminants.

In a vast region that is difficult to access, such as the Arctic, it is not easy to predict what the future will be with regard to contaminants. Most likely, the persistent organic pollutants will decrease, but throughout this century we will still have to face negative impacts. Contaminants in marine food is not as problematic around Svalbard as it is for Greenland and Arctic Canada, where indigenous people obtain much of their subsistence from higher marine trophic levels. Currently, the levels of oil-related contaminants are very low in the Arctic Ocean. With oil and gas exploration and production in hitherto ice-covered regions on the rise, and the increasing amount of traffic generated by tourism and commercial vessels bound to and from Asia, the amounts of these contaminants will sharply increase in the future. In addition, on a local scale, effluents from mines may create pollution.

Arctic organisms, in particular phytoplankton, are at risk of being damaged by ultraviolet radiation. The concentration of ozone-killing chemicals has decreased in the Arctic atmosphere, but the ozone concentration is not expected to remain at the same levels as before the mid-20th century. Special atmospheric conditions may result in deepening of the ozone hole and its expansion southwards. Meanwhile, small organisms have to cope with the negative impact that the radiation has on their metabolism.

In essence, while major contaminants decline, new long-transported ones are steadily supplied to the Arctic along with the revolatilisation from local sources. Oil and gas pollution will increase rapidly in the near future, in response to commercial exploration, exploitation, and traffic increases. ∎

Sea ice decline is a threat to species

Sea-ice extent and sea-ice thickness have decreased over the last 30 years, with a record-low ice cover since 2007 and a new minimum in 2012. This has consequences for the warming of the Arctic Ocean, and for species that depend on sea ice as a substrate. Sea-ice extent is declining more rapidly than models have predicted and it may prove to be the case that the North Pole and most of the Arctic Ocean will become a more or less ice-free region by early autumn each year, already before mid-21st century. The self-amplification of ocean warming that follows the exposure of seawater (which is dark compared to ice) is of major concern. Melting sea ice also changes biological diversity. Several key organisms in the Arctic Ocean ecosystem, e.g. sea-ice algae that grow in and under the ice, will suffer loss of habitat. Enigmatic Arctic fauna, such as seals and polar bears, may also become deprived of an essential substrate. Some seals give birth to their offspring or moult on ice. Polar bears depend on seals for food and the marginal ice zone is their natural hunting ground. Whereas the former species may survive on land (from where it developed 1.5 million years ago (?), with periodic cross-breeding with its close relative, the brown bear), the seals may encounter greater difficulties. The lifecycle of some other organisms, including several species of seabirds, is connected to sea ice.

Zooplankton may depend upon sea-ice algae early in the productive season and they, in turn, can be essential food for key Arctic fish species. To obtain the right type of food, in sufficient amounts, at the right time depends on suitable production conditions along the marginal ice zone.

Sea-ice melting has also resulted in sea-ice thinning, with more than 50% of the volume lost since the start of thorough registration round 1979. Thick ice is not annual, but multiyear ice (annual ice that has been formed during wintering inside the Arctic Ocean). The predominating annual ice melts quicker than multiyear ice and promotes a reduction in sea-ice cover. The consequences of increased amounts of fresh-water for the future development of the Arctic Ocean, its productivity, and the global climate, are currently debated, but an ice-free period at the end of summer as well as strongly stratified surface water will occur soon.

In essence, global warming will cause the summer sea ice to disappear in the near future, opening up for an ice-free Arctic Ocean. This will be an immense human impact, the ecological and biogeochemical consequences which can hardly be estimated. ∎

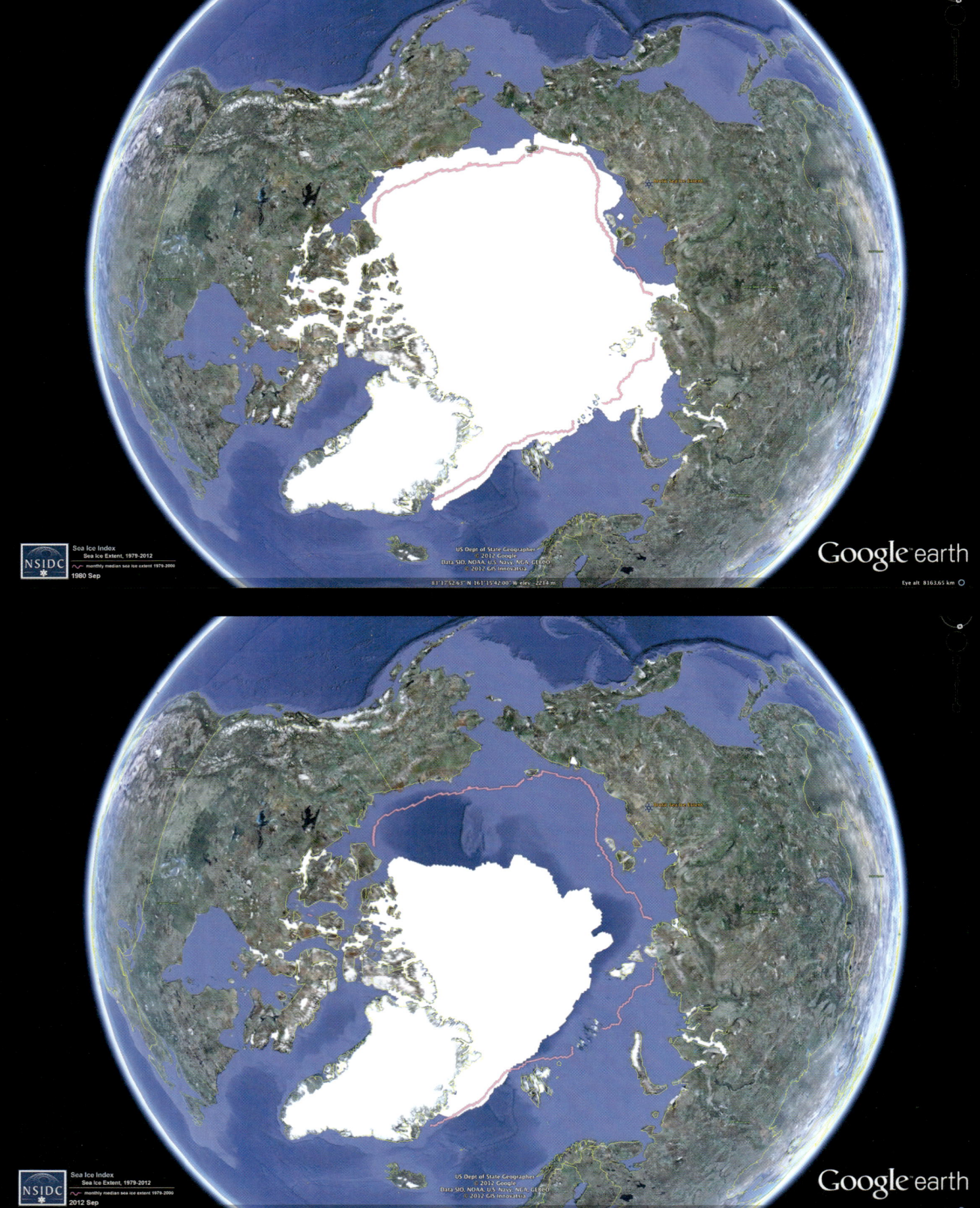

A warmer ocean changes ecosystems

A warmer Arctic Ocean will result in ecosystem changes. Warmer water will result in less ice, and thus more radiation will be available for plant growth. Primary production will increase until limited by nutrients. However, increased temperatures also imply that respiration of both plants and animals increases. Increases in temperature cause much higher increases in respiration than in primary production. Consequently, the food available for higher organisms may decrease as water temperatures increase. However, with increasing temperature, species that prefer warmer waters may be able to move northwards as their ecological niche widens. Indeed, several southern species are reported to have spread northwards already to Svalbard. Thus, both the basic ecosystem function and the composition of species will change in a warming Arctic Ocean.

Over the last 100 years, northwards movements of warmer water species have been observed in western Spitsbergen, at the expense of Arctic species. The 'battle' between warmer and colder water species is continuous and the regions where Atlantic water spreads north along the west coast of Spitsbergen constitute one of the major battlegrounds. An analysis of a 30-year time series of bottom flora and fauna by the University of Tromsø revealed that climate warming undoubtedly could trigger abrupt ecosystem changes in the Arctic. The most striking component of the benthic reorganisation was an abrupt fivefold increase in macroalgal cover in 1995 in Kongsfjorden and an eightfold increase in 2000 in Smeerenburgfjorden. The abrupt, substantial, and persistent nature of the changes observed is indicative of a climate-driven ecological regime shift. The ecological processes thought to drive the observed regime shifts are likely to promote the borealisation of these Arctic marine communities in the coming years.

Essential key organisms for fish and birds in the waters around Svalbard are the two most important zooplankton species, namely Arctic and Atlantic copepods (the adults of which grow to 1-3 millimetres in length). The success rate of their reproduction and survival depends on water temperature. As Arctic copepods contain far more fat then Atlantic copepods, and as fat is the key to life everywhere in the Arctic, warmer waters have a profound impact upon the energy flow. Commercially significant fish species such as cod and capelin may change their distribution and spread from today's region in the southern to central Barents Sea and into a northerly or north-easterly direction, with potentially negative consequences for Norwegian fish quotas.

The breeding cycle of some copepods, shellfish, and fish coincides with the spring bloom, which enhances their survival. Shifts in the timing and spatial distribution of seasonal production could disrupt the corresponding lifecycle events, thereby leading to decreased survival. In addition, the loss of sea ice in the summer months is expected to enhance secondary pelagic production, with associated changes in the energy pathways within the marine ecosystems. These changes are expected to alter species composition and the carrying capacity of marine habitats, with associated impacts on the ability of the region to support marine fish and shellfish populations.

The marine ecosystems around Svalbard have developed since the last glacial period and are adapted to natural climate variability. The marine waters around Svalbard have developed a considerable ability to endure changes before drastic alterations will occur. If temperatures increase beyond the natural variability of today's climate, the uncertainty connected to the response of ecosystems and their resilience will increase too. Although climate is probably the most important factor that determines ecosystems, there are several other influential factors, such as invasive species and plasticity in physiological response. The sum of these factors will determine the outcome.

Graphs of sea-ice cover recorded in March (maximum) and September (minimum) (National Ice and Snow Data Centre, Colorado, USA) are shown on page 294. The upper graphs indicate sea-ice cover in March (left: 1979 (at the start of remote sensing of sea-ice by satellite); right: 2012. The difference between the graphs is small, i.e. climate change has not resulted in major changes in the winter-ice distribution (but the ice is much thinner today compared to 30 years ago). The lower graphs indicate sea-ice cover in September (left: 1979; right: 2012). The rapid decline of ice at the end of summer, caused by global warming, is clear: there is far less summer ice now than 30 years ago.

In essence, warming of the Arctic Ocean will result in rapid changes in marine ecosystems and increased production of algae, but in some regions also a decreased production of key zooplankton species. In some regions around Svalbard and parts of the Barents Sea, the general productivity will rather decrease than increase. ■

The inescapable acidification of the ocean

The steady increase in atmospheric carbon dioxide concentration results not only in global warming but also in acidification of the ocean, which is strongly buffered against such changes. Until recently, the ocean has absorbed more then 25% of the carbon dioxide that humans have produced since the Industrial Revolution, without any major increases in acidification. As the ocean has already taken up as much as possible, the buffer capacity is approaching a limit, at which point acidification will commence. All available models for the present century predict that the atmospheric carbon dioxide concentrations will increase steadily. They will reach levels that have not prevailed on earth for the last 20 million years. The Arctic Ocean will become the first of the world's oceans to be affected by acidification. As the water chemistry changes, certain organisms with calcareous shells will face difficulties by the end of the century, including deep-water coral reefs, some plankton organisms, and shells. Dissolution of shells can already be observed today. In the long run, either such organisms will perish, their numbers will decline, or they will find ways to protect themselves.

The production of a calcareous shell is energy-demanding. As the outer surfaces of the shell dissolves, gradually more and more energy has to be devoted to the production of new shell. There are two different types of calcareous shells: those that consist of calcite and those that consist of aragonite. Small, winged snails, an important component of Arctic plankton, form shells of aragonite that dissolves more rapidly than calcite. Thus, the winged snails probably will be the first group of organisms to be affected by acidification. Another group of organisms that could be seriously affected are deep-water corals, although these are less common around the Svalbard archipelago compared to the Norwegian coast. Generally speaking, young, shell-bearing organisms that live at the bottom of the ocean, are particularly sensitive to acidification.

A pteropod (from the Greek, meaning 'wing-foot'), is shown on page 297. It is a term applied to groups of specialised free-swimming pelagic sea snails and sea slugs. The term Pteropoda no longer has a precise scientific use, but the vernacular name pteropods is still sometimes used for convenience. Also shown is a sea butterfly (*Limacina helicina*), which has a shell that will easily be weakened by the acidified Arctic Ocean of tomorrow.

In essence, acidification has already begun in the Arctic Ocean and in decades to come it will change the composition of ecosystems. It will not be possible for this trend to be reverted in the foreseeable future, because it reflects carbon dioxide emissions since the Industrial Revolution, which have depleted the ocean's buffering capacity. ■

Landscapes and animals

Climate determines the distribution and lifecycles of plants and animals. Annual maximal and minimal temperatures are one of the forcing elements that determine the regions in which organisms can exist. The length of their productive period is another important factor. For example, if we assume an active growth season of two months for plants in Svalbard, an earlier start and later end to the productive season by two weeks in each case, would increase the productive period by 50%. A warmer climate would result in displacement of Arctic species northwards or to higher elevations, whereas species from warmer regions would invade from the south. The enigmatic terrestrial mammal in Svalbard, the Svalbard reindeer, would benefit from more plant growth and a longer productive season. However, more winters with ice-covered snow after rainy periods could have a negative influence on the population. Increased summer temperature would also result in more insect attacks and parasites. So far, the population has increased, indicating a positive response to warming.

New terrestrial species must have the ability to spread to Svalbard from either the Fennoscandic peninsula or Novaja Zemlja. Plants, microorganisms, invertebrates, and birds are able to spread to isolated islands, but terrestrial mammals face greater difficulties. This protects the terrestrial mammals in Svalbard from rivalry from competitors from the continent. The closest migration route would be from the east and Novaja Zemlja, but less ice cover also in the eastern Barents Sea makes this connection unlikely. Only one new species has arrived so far, involuntarily introduced from Siberia: the southern vole, which can be found in the deserted Russian mining town Grumantbyen, in Isfjorden.

In essence, Svalbard's landscapes will change as the vegetative period lengthens and the summer temperatures rise, but new invading terrestrial mammals are not expected. ∎

A passion for pristine environments:

Arctic tourism

Tourism in the Arctic is influenced by climate change in several ways. In Svalbard, tourism is one of the main businesses. It has increased significantly as a result of advertising, better infrastructure, less ice, and easier access to places that normally are difficult to access. In addition, increased environmental awareness has led people to pay attention to pristine nature. Carbon emissions in Svalbard account for approximately 1% of the total Norwegian emissions. However, the figure is almost 12 times more per capita than in Norway. Most of the emissions come from tourism, followed by coal-based energy production and from ships: in the period 2000–2007, ship-based emissions increased by approximately 30%, but the emissions from ships carrying tourists increased by 70%. Tourism activities that have caused landfalls have already resulted in damage to vegetation and historic remains and objects. It is thus evident that ecotourism is one of the main contributors to carbon emissions and terrestrial ecosystem damage in Svalbard. Further, more and more tourists go ashore at an increasing number of places in Svalbard. The government has already introduced restrictions, and the eastern part of Svalbard is now closed to tourists. In the near future, ships transporting heavy oil will not be allowed to enter Svalbard waters, reducing not only the potential damage in the case of accidents, but also the number of ships that can operate there. In addition, it is now mandatory for cruise vessels to have a Norwegian pilot. This can create bottle necks that will further reduce ship-based tourism in Svalbard. All of the above changes will probably have negative consequences for the tourism industry. In order to preserve the pristine nature of Svalbard, Norway will have to define the upper limits of mining, inhabitants, and research in the near future. Otherwise, it will have to learn how to store or remove CO_2 emissions.

The tourism industry in Svalbard is strongly dependent upon the archipelago's wild and pristine nature, and has to evaluate carefully the applied logistics of visits and the general ecological situation. The genuine interests of humans in ecology, the excitement of travelling to remote locations and enjoying wilderness are a far cry from everyday life. However, such tourism should not be allowed to increase the velocity of ecosystem degradation. The foundation of ecotourism should be ecosystem sustainability. If that cannot be ensured, tourism as such or the manner in which it is practised will have to be re-evaluated.

Inspired by an uncorrupted civilisation and armed with high-power binoculars, tourists have spotted something on the horizon. Is it a polar bear, a whale, or merely a seal? Such touristic enthusiasm has not changed over time. On page 300, a photograph taken in Svalbard in 1896 or 1897, by the painter Hans Wieland, shows four gentlemen who are apparently extremely enthusiastic about the sights.

In essence, the pressure of tourism on Svalbard's ecology has significantly increased, but steps have been taken to control and limit the impact. It is in the long-term interests of tourists, Svalbard's inhabitants, and humanity to limit the human impact on Arctic nature and reach a more sustainable utilisation. ∎

Infrastructure challenges

To maintain societal performance, infrastructure (e.g. buildings, roads, airports, harbours, power plants, power lines, and water supplies) must be continuously functional, and this has implications for constructions, the selection of materials, design, and localisation. In many circumstances, the largest infrastructure damage is instigated through a combination of various climatic challenges. In the High North, examples include intense precipitation and strong winds, precipitation at temperatures around zero degrees (ice rain), and sudden temperature increases (leading to floods). Good emergency preparedness for all infrastructure aspects is needed to ensure perfect collaboration, the base of a well-functioning society.

The current infrastructure in Svalbard has to take climate change into consideration. Wetter weather, freezing rain, avalanches, and early thawing may result in disruptions to traffic, particularly air traffic to and within Svalbard. The reduction in permafrost will have a negative impact upon existing houses, roads, and airport runways. High air temperatures during summer may prove uncomfortable in well-insulated houses. Extreme weather, such as heavy precipitation, may challenge the current drainage system and flooding may occur. The shrinking of glaciers may expose settlements such as Longyearbyen to reduced drinking water supplies. The increase in seawater levels, of major concern in many regions of the World, is not a major problem for Svalbard, due to the rocky and steep nature of most of the fjords.

In essence, decreased infrastructure vulnerability requires that society thinks ahead, but that may imply far more costly solutions, and may demand new designs or even new locations. There is a need for more emphasis on cold climate technology. For Svalbard as a whole, solutions that are more robust have to be planned and investments made. ■

Adaptation to change*

Warming and cooling in the Arctic Ocean have taken place over much longer time periods compared to the rapid temperature increases in recent decades and anticipated in forthcoming decades. Adaptation mechanisms that have been adequate to date may be insufficient in the future. One may talk about dangerous climate change that has been defined as events precluding ecosystem adaption, jeopardizing food production, and preventing sustainable development. In the Arctic, the current rate of climate change appears faster than ecosystems can adapt naturally, and Inuit communities are already experiencing compromised food security and health, and threats to their traditional cultural activities.

Society has to and will adapt to change, particularly climate change. If adaptation does not take place voluntarily, realities will force an unprepared society to change: 'Those that are late are punished by life', said Mikhail Gorbachev just before the Berlin Wall fell. To discover the signs of change early enough is a precondition for any society or ultimately humanity to maintain a high standard of living. This is the foundation for the lively debate between scientists, the United Nations Intergovernmental Panel on Climate Change, and environmentalists on the one, and politicians, nations, and humanity on the other hand – often termed the climate debate. Will the first group take the good life from humankind or is the second group neglecting the signs of change (and hence also taking the good life away from humankind)? Should we believe in climate change and prepare for it? Climatic and other large-scale changes can have potentially large effects on Arctic communities, where relatively small and narrow-based economies leave a narrower range of adaptive choices.

In Svalbard, decisions have to be made as to how much ship traffic is permissible for touristic purposes, what tourists may be allowed to do upon their arrival, and what the upper limit of the tourism industry should be. The same also applies to scientific activities and the numbers of people permitted to live in Svalbard (and what they will be allowed to do while living there). The air quality in Longyearbyen can be as bad as in many central European cities at the height of the scooter season. How many scooters should be allowed and much fuel should each scooter be allowed to consume during a year? Should energy production be based upon coal? Could one use sources that have lower carbon dioxide and soot emission? Is it possible to store CO_2?

In order to keep Svalbard society functioning well for the foreseeable future, all new infrastructure needs to have dimensions that take care of the changing climate. Also, the fishing industry in the north has to consider the climatic challenges ahead. Fisheries may change due to invading warm water species. Harvestable production is likely to increase in the northern Barents Sea, but may decrease in the south, where most of today's fishing grounds are located. The energy flow through the marine food web changes and some valuable fish resources may shift out of Norwegian territorial waters.

Currently, fishermen and reindeer herders often do not take climate change particularly seriously. Why is this so? They need always to adapt to changes and do not question in too much detail what the long-term future may bring. Rather, it is always the forthcoming year that deserves their continuous attention. Also, the greatest adaptation challenges do not derive from nature, but from ministries and other political bodies that determine quotas and regulations (such as taxation, customs barriers, export regulations, and veterinary rules). Thus, climate change is not at the forefront of fishermen's and reindeer herders' minds. Yet, in reality, both groups champion the art of year-to-year adaptation.

In essence, it is essential to identify how vulnerable Svalbard society is regarding climate change, in order to develop effective adaptation measures. Not doing so would leave any society at risk of developing adversely. Most scientists believe that this type of risk is already high and that it will increase. ∎

---

* 'The great French writer Victor Hugo wrote that there exists nothing more powerful than an idea which time has come. I believe that also the opposite is true: that there is nothing more destructive than an idea which time has come but which humans refuse to accept' (B. Clinton, in Rockström & Klum, 2012)

X: OUTLOOK AND PERSPECTIVE

HUMAN DEVELOPMENT AND ECOSYSTEM CHANGE
MYTHS ABOUT THE CLIMATE
CHANGE TO NEW EQUILIBRIUM?
COMMUNICATE WITH AND TO THE PEOPLE?
THE PEOPLES OF THE ARCTIC
THE PEOPLE OF THE ARCTIC
SOCIALISING LOSSES, PRIVATISING PROFIT
EXPLOITATION, PRESERVATION AND SUSTAINABILITY?
RESILIENCE ASSESSMENTS AND TIPPING POINTS
MULTIDISCIPLINARY AWARENESS: A PRECONDITION FOR RESILIENCE MANAGEMENT
THE BEGINNING OF THE END?
CONFIDENCE IN THE FUTURE: A LOST SKILL?
THE SERENITY OF THE MIND

Human development and ecosystem change*

Humanity is presently traumatised by worries about climate change. The sea level is rising, and sea ice, glaciers, and permafrost are gradually melting. Charismatic species such as the polar bear may disappear. All this is caused by increasing global temperatures. Thus, a major question arises: Who turned up the heat? The answer to this question is heavily debated in the media worldwide.

The notorious misunderstanding of the basic terms *weather* and *climate* (which is the average weather of the last 30 years) greatly complicates people's understanding of climate change. The general public often interprets natural variability in weather as climate change. Who has an exact memory of the weather over a 30-year period? To register climate change, statistics have to be applied. Based upon evidence generated by such statistics, there is currently international consensus among scientists that climate change happens and that it is *very likely* that increased emissions of climate gasses have caused most of the increase in global temperature during the last 50 years. Since the start of the Industrial Revolution (about 1750), carbon dioxide (caused by, e.g. burning oil, coal, and gas, deforestation, and drainage of wetland) and methane (caused by e.g. the burning of oil, coal, gas, refuse tips, ruminants - especially cattle, rice, and melting permafrost) have increased with more than 35% and 150% in the atmosphere, respectively. This implies that we, the protagonists of the Industrial Revolution and concomitant time period, have turned up the heat. The climate of the earth is far from what could be expected within the range of natural variability. This scientific fact is frequently vigorously rejected by variable fractions of the world population.

Nowhere on Earth has atmospheric temperature increased more than in the Arctic, and it will steadily increase, also in the future. The Arctic ecosystems have changed already and will continue doing so in this century. They will reach a state that has not been experienced for tens of thousands of years, and possibly millions. Scientists are not only concerned about the absolute change in temperature, but also the *pace of change* in the Arctic. The expected environmental consequences are extraordinary and mind-boggling.

In principle, this is nothing new under the sun! Humans have impacted biodiversity and ecosystems throughout history. From the earliest *Homo sapiens* at the hunter-gatherer stage, when only a few million individuals existed, hunting accounted for significant biodiversity losses, such as the disappearance of charismatic mammals, e.g. mammoths and sable tigers. Later, the blossoming cultures of the Fertile Crescent gave way to infertile steppes and deserts due to human action. In the Middle Ages and later, the deforestation of Central Europe was probably the greatest environmental change ever to have happened in the region. More recently, the resource-exploiting strategies of the Industrial Revolution have become global and impacted the earth's climate and ecosystems irreversibly. We live in a shrinking world. We live in the time of the Anthropocene, the era in which human activity influences the biogeochemical cycling of the entire earth. Today, here we are, living on our one planet, with serious limitations ranging far and wide, and a strategy that neither supports sustainability nor strengthens resilience – here, defined as the capacity of a system to absorb disturbance and reorganise in order to retain essentially the same function, structure, and feedbacks of the system. ∎

---

* 'In my own view, it is only a narrow passage of truth (regardless of whether it is a scientific or other truth) that passes between the Scylla of a blue fog of mysticism or a Charybdis of a sterile rationalism. This will always be full of pitfalls and one can fall down either side' (W. Pauli, 1952)

Myths about the climate*

To obtain one all-embracing depiction of the world has always been a dream of humans. Today's dream is to explain the entire world through natural science. At any time, such dreams result in the repression of knowledge that is considered inadequate and does not support the right path. In other words, an ideologization of otherwise good concepts takes place. For example, it took considerable time before it became evident that the founder of modern science, Sir Isaac Newton (1642–1727), was deeply dedicated to calculating how big the New Jerusalem had to be to receive all souls at Judgment Day. He was much more into alchemy than physics. He told a friend that he was less concerned with physics, which he worked on in his spare time. John Maynard Keynes claims that 'Newton was not the first of the age of reason. He was the last magician'. This is also true of another protagonist of the modern age, the co-founder of modern astronomy, Johannes Kepler (1571–1630). When do we read such evidence in today's books on physics? Rationalism and the dominance of natural science attempt to put the world into straightjackets, and exclude all knowledge that does not fit. Humans do not like straightjackets. They like to keep a balance between the various claims to understand the world. As a reaction to the straightjackets that all ideologies provide, humans are attentive to a variety of beliefs. They try to construct mosaic-like, holistic worldviews. Thus, suites of wise or devious myths are part of our existence. Thus, too, the resistance against the fact that climate change is caused by man is to a large extent caused by the attitude of natural scientists themselves.

In the following, I scrutinise a few common myths that deal with the present-day or future climate.

'It is not dangerous if global temperatures increase by a few degrees.'
This is true, but the effects are unevenly distributed. They will hit the low-populated Arctic and some high-populated regions in the south hard, while a few (and often rich) countries will not suffer much. The climate challenge infiltrates a suite of human problems, such as health, political stability, economic strength, and social security. This has consequences for all countries and their citizens. For example, affluent nations need to prepare for hundreds of millions of climate refugees, but are they doing so?

'It has been warm also in the past, particularly in the Arctic. The climate changes, whatever we do. Why should this be different today?'
The answer is, because this time the warming is caused by humans (we should instead be in a cooling phase). We know that our lifestyles have turned on the heat and this will change our lives and livelihood. Humankind probably does not fully comprehend the consequences of the pathways that have been selected.

'Climate researchers and ecologists just quarrel – they cannot agree on anything. Basically, we know too little about the climate.'
Most of the quarrelling does not take place between active scientists, but between scientists and either laypersons or politicians, often involved by media that wish to expose a 'balanced' debate. There can be disagreement and there is, of course, uncertainty, but the media-centred climate debates, stimulated by our endless desire to be notified about records and weather calamities, are not much visible in international journals and at conferences. The United Nations Intergovernmental Panel on Climate Change reports are based upon international consensus and thus soundly non-radical and conservative.

'Everything happening to the climate is caused by humans.'
This is certainly not the case.

'Nothing happening to the climate is caused by humans.'
Our most advanced methods for dealing with the development of many types of observations in space and time and models conclude that this assumption is wrong.

Will we be able to leave the era of myths? I believe that we can never leave. Insight and wisdom are one of man's best qualities for keeping chaos at a distance, but it is only one of several merits of the human psyche. The question is rather how do we deal with our indispensible myths? To what extent are we able and willing to shed rational light on the gloom that surrounds myths? Myths tell us something about our inadequacies; they are our friends that come bearing a message. Regarding the myths mentioned above, it is time to scrutinise them and enter a new era with a cool, rational mind. Will natural scientists and media support humans in these challenging times through *weighted* contributions (knowing what they write about), not just *balanced* contributions (contrasting opinions)? Alternatively, will we continue with the myths that prevent us from real comprehension, the base from which to turn knowledge into action? ∎

---

* 'If your face is crooked, don't blame the mirror'
(N. Gogol, 1842)

Change to a new equilibrium?

The main goal of the United Nations Framework Convention on Climate Change is to stabilise concentrations of greenhouse gasses in the atmosphere at a level that will prevent dangerous man-made impacts on the climate system. Scientific evaluation indicates that seriously negative effects of climate change occur above 2°C atmospheric warming (global average). Consequently, the European Union and several other nations declared that the average surface temperature of the earth in 2100 should not rise by more than 2°C, compared to pre-industrial times. Today's growth in population in combination with agricultural and industrial schemes implies that this goal is already out of reach. Our best global temperature estimates for the end of the century are closer to +4°C. Several Arctic-tipping elements have been identified that will result in dangerous change by the end of the 21st century (e.g. sea-ice cover and thickness, the Greenland ice cover, permafrost, and boreal forests), thereby endangering people's livelihoods. However, we seem too paralysed or not willing to admit the problem and take action.

The quantities of climate gasses that have entered or are steadily being released into the atmosphere have a long-lasting effect on our climate. Even if we were to cut back strongly now, the warming effect would continue for decades. There is no way back in the short run. Humanity has to face what has been done thus far, but may have the wisdom to look to the generations ahead of us. That was the norm in preindustrial times, which were characterised by longer time perspectives. Limiting growth for the benefit of one's grandchildren and great-grandchildren is not a new strategy: it is an old-established, wise, and conscientious one that the industrial nations have lost out of sight and have exchanged for a new credo: growth, at all costs.

By the end of this century, after passing several tipping points, the Arctic Ocean ecosystems may have shifted towards a new ecological steady state, accompanied by not fully understood consequences. With regard to species, systems, and productivity, no one fully comprehends the necessary details that this state implies. Until then, we can expect major and rapid changes, ecosystem shifts, and low predictability. We are witnessing an experiment where even the best Arctic scientist has rather to observe change than to predict it with precision. The setting thus exposes scientists to a sense of serious unease. ∎

Communicate with and to the people?

'More research is needed' is a common demand when confronted with a dilemma. However, the world does not need more research to make the decisions to slow down the deterioration of the Earth System and to limit the negative consequences that lie ahead of us. Research can be used to postpone political action, and researchers can suffer from the ethical dilemma that derives from having their work misused in order to postpone political decisions. How can scientists communicate their knowledge in order to influence political decisions and improve the world's managements regarding climate change?

Scientists, like other professionals, have their share in the exclusion, disregarding, and rejection of knowledge. Insight is often suppressed if the consequences are undesirable and/or unacceptable, and knowledge can be rejected when connected to 'fear messages'. The representatives of science may have to look at how they present 'the truth' to society and politicians. Should communication to the general public be based upon the necessary objectivity and psychological distance that characterises science? Must science be communicated in an authoritarian manner, ex cathedra? At times, science bears a resemblance to a kind of superego that in a cruel manner bombards us with impossible demands and then gleefully observes our failure when we do not meet them. One can raise doubts about whether this is a strategy that will induce a deeper understanding and reach the hearts of humans. Do we need alternative ways of communication, where scientist share their knowledge with the general public in a non-alienating and inter alia manner? Obtaining knowledge in a strict scientific manner and reaching out to the general public appear to be different roles that need discrete strategies. Scientists are specialists in knowledge communication to their own kind, but with regard to the general public they are often well meaning, but amateurs. ■

The peoples of the Arctic

The peoples of the Arctic are its indigenous inhabitants, consisting of small, scattered groups across the pan-Arctic realm. One of these groups, the Inuits, sustain themselves primarily through marine food sources such as seals, fish, and birds (and additional terrestrial supplies). There is a wide range of groups that traditionally have mainly practised reindeer herding, such as the Sami, Chukchi, Koyaki, Yakuts, and Nenets.

The most compelling characteristic of the peoples of the Arctic is that they master the art of adaptation when it comes to climate and ecosystems. The Arctic ecosystems are strongly forced by climate, and the peoples of the Arctic have learned to adapt accordingly. Southerners often focus upon the negative consequences of climate change for the peoples in the Arctic. Northerners seem far more concerned about unpredictable state-controlled management and departmental directives than climate change. Decisions by institutions, situated in capitals far to the south, have often serious and at times detrimental consequences for the livelihoods of the peoples living in the Arctic. The embargo on seal products imposed by the European Union is one example. This decision does not protect Arctic seals (the populations of which have almost reached the highest levels since estimates have been available), but the decision deprives Inuit hunters of one of their few ways to earn money. Money is a necessity, even in the most remote regions where subsistence is based upon hunting. Increasing seal populations add additional pressure on fish resources throughout the Arctic and fishery has by far the strongest bearing on the Greenlandic economy. An attempt to prohibit the traditional castration of reindeer by herders is a second example. It represented a serious threat to reindeer herding in Norway, but was fortunately defeated. The difference between raising cows or reindeer and the cultural implications may be difficult to grasp for governments' agricultural management institutions. A third example is the manner in which mandatory school education was introduced to the peoples of the Arctic. At times, it had seriously negative consequences for peoples that move continuously or inhabit scattered settlements. It resulted in larger, permanent settlements far away from the hunting grounds or pastures, depriving youths of cultural connections and even removing them from their own culture. Alternatively, it resulted in alienation, and in cases where native languages were forbidden at boarding schools, often in the opposite of what was intended, namely illiteracy. It is indeed favourable that the majority and those that have the power are sensitive and respectful when it comes to the cultural distinctions of ethnic minorities.

The increase in infrastructure in Arctic regions is another threat to indigenous people. Roads, military bases, shipping, fishing, mining, and oil/gas extraction are examples where the still reasonably pristine Arctic region is invaded in order to satisfy the needs for control, power, and resources by the non-Arctic majority. The Arctic is no longer north of the last frontier. The frontier has been pushed rapidly northwards; has already entered the central Arctic Ocean and will soon dissolve. To make a traditional living amid roads, military townships, mines, aluminium plants, and oil/gas towers is an art that the many indigenous Arctic people have had to develop. In several cases, their traditional existence is threatened.

Sooner or later, all indigenous people of the Arctic will face the hardship of climate warming, which is so rapid that they hardly will be prepared for the pace. It will first of all appear through direct ecosystem impact and industrial and infrastructure development. Life will, or already has been, turned upside-down. As long the peoples of the Arctic can recall, they have been exposed to frozen water. In the future, they will be subjected to the opposite: melted ice. ■

The people of the Arctic

The above heading is not a spelling mistake: the people of the Arctic are all people that live in the Arctic. The indigenous people, the peoples of the Arctic, account for only approximately 10% of the 4–10 million people that currently live in the Arctic (depending on geographic definitions). Here, I wish to focus upon this large and frequently not appropriately addressed majority of people in the Arctic realm. Many Norwegian, Russian, Canadian, and US citizens visited the Arctic 100–200 years or longer ago. The descendants of these groups cannot claim any of the rights provided to or claimed by indigenous people, yet are they strangers – southern intruders? Are they citizens that look at the High North as the last frontier to gain rapid wealth and leave after making a profit? No, the High North is, along with the indigenous people, also their homeland. Their voices, modulated by living alongside indigenous people, mixed marriages, shaped by Arctic realities, and empathic consideration of a joint historic cultural repression, must also be heard.

Most of the non-indigenous people in the north live in the Arctic due to resource extractions (fisheries, agriculture, mining, and oil/gas), infrastructure, tourism, and military and administrative duties for their respective national states. Over lengthy periods of time, they have built up a close relationship with their environment. They care for the future of all in the High North. Living in the Arctic is strongly resource-dependent and thus all of its inhabitants care for the future of existing supplies. There has been a clear tendency by the majority of industrialised nations to venture into the Arctic, investigate the resources, extract them, and leave after the exploitation. This is detrimental for the people of the Arctic, whose future is based upon sustainability of the precious resources. There is therefore not only the traditional conflict between indigenous and non-indigenous people in the High North, but also between southerners and northerners, generally speaking. The challenge of the future is to overcome the internal indifferences between the people and the peoples of the Arctic, and to develop a unified position against those that wish to exploit natural resources in a non-sustainable manner, move the revenue to other markets, and then leave a damaged environment behind, with little local capital or expertise left in a pauperised region. In order to be successful in such a challenging setting, one has to promote the level of education and strength of the political institutions of the people of the Arctic. This is an important precondition to strengthen the capacity of the population to tackle challenges, enabling them to defend their local social-economic system and shape their future. ∎

Socialising losses, privatising profit

As mentioned earlier, the general attitude of the majority of nations regarding the Arctic is embodied in the metaphor 'the last frontier'. The High North seemingly bears similarities to the Third World. It provides affluent parts of the world with raw materials, but neither controls resource exploitation nor enjoys the benefits of processing raw materials into finished goods. The availability of resources beyond the last frontier is explored and extracted by nation states that most often are the owners of the 'no-man's-land' of Arctic regions. Consequently, environmental impact assessments and resource access are regulated through the respective, central governments. Local rights to resources, let alone private ownership, are not the rule in the Arctic. Companies, some even owned by the states (i.e. all citizens), extract the resources, which are considered national because they are owned by the states. This implies that the profit basically is privatised (with considerable profits to the nation states and hence to the majority of the population), but the eventual (most of the time inevitable) environmental damage and ecological losses are socialised locally. Initially, they become the responsibility of the affected local communities, but eventually also of the entire population of national states, if it cares for 'so few'. The majority of the inhabitants of the different states live in the south, far away from the damage, and consider environmental damage often insignificant when related to the interests and benefits of their nation as a whole. In Svalbard, examples include coal mining (now), the depletion of whales and walruses (in the past), and the depletion of oil and gas (in the future), but in contrast to other parts of the Arctic no lifecycle society is present in Svalbard. The degradation of ecosystems and the loss of biodiversity are consequences for the local or adjacent societies, not the beneficiaries of the business that profit from the archipelago's resources.

The revenue most often passes southward, where the capitals and financial centres are located and most people live. **The disruption and mess** is left behind for the people of the Arctic to cope with. They deal with both the short, lucrative exploitation period and the long time phase of negative environmental impact. Some of the revenue is, of course, returned in the form of subsidies and infrastructure, but the prevailing main impression is that the south subsidises the north, not that the north supplies the south with significant resources to earn money. This supports the notion of an underdeveloped, support-dependent north and a sense of economic and cultural superiority among the majority of people in nations that have Arctic territories. Can we detect remnants of long-forgotten colonialism and the 'white man's burden'?

When will resource exploitation include the true cost of environmental destruction and the recovery of nature? While the Arctic, generally speaking, is still regarded as pristine and thus comprises all of its natural value, the real price of resource exploitation needs to be approximated. Furthermore, steps have to be taken to ensure that more than short-term profits or long-term losses are left to the locals. In the future, not only the profits but also the losses will have to be 'privatised'. Also, greater respect has to be paid to the people of the Arctic and their dignity, to prepare for a wonderful future of their own, not merely a majority-imposed future. ■

Exploitation, conservation and sustainability?

There are generic debates in Arctic and adjacent nations that reflect strong differences in opinion regarding the Arctic region. The increasing lack of resources for the steadily growing world economy should result in new endeavours to make new resources available, say economists, industry managers and many politicians, who are concerned about the future development. Seriously concerned about the future of our planet, the pristine nature of the Arctic (the result of ice cover, an inhospitable environment, technologic challenges, and the remoteness) should be preserved, say 'green' movements or environmental ministers. The people of the north have a double agenda. They do not wish to open a region to further exploitation that privatises the profits in the hands of the few and socialises the environmental losses to a local majority. However, they do not want their homeland transformed into a natural park either, one that preserves not only nature, but also their lives and future. Who wishes for all time to live in a museum, even if it is an extraordinary one? These two positions have given rise to profound internal debates.

Some wish to give specific rights only to the vulnerable and easily marginalised indigenous people, such as subsistence hunting of marine mammals and mineral resource extraction. The people of the north – in concert – should argue for the viewpoint that the Arctic is recognised as their homeland. It implies that it should provide them with undisputable rights to exploit and manage the resources. However, they need also to take full responsibility for the sustainability of their homeland. The conflict between preservation and exploitation, frequently proceeding without the people of the North's involvement, can only be resolved and turned into a fruitful dialogue through the concept of Earth System Science and resilience, two major modern concepts that seem essential to tackle the challenges of our age, the Anthropocene. Can exploitation and conservation be combined? Further, can all fractions of people in nations be involved and influence the outcome?

Earth System Science embraces chemistry, physics, biology, mathematics, and applied sciences in transcending disciplinary boundaries to treat the earth as an integrated system and seek a deeper understanding of the physical, chemical, biological, and human interactions that have determined the past, current, and future states of the earth. Earth System Science provides a physical basis for understanding the world in which we live and in which humankind seeks to achieve sustainability. Sustainability, however, rests upon resilience (i.e. the capacity of a system to absorb disturbance and reorganise in order to retain essentially the same system). The concept of resilience includes the natural environment, such as ecosystems and social environments (e.g. people, cultures, governance arrangements, and economic structures), and the links between them. Social changes interact with changes in the physical environment and ecosystems. The resilience of a system is a measure of its ability to respond to changes and shocks while keeping its identity. Resilience thinking involves *both* a given ecosystem *and* the humans it supports. The goal is to reduce the pressure on a system sufficiently early to avoid tipping points or points of no return. How can the Arctic and its communities absorb disturbances and maintain function? How big is the buffering capacity of a system? What are the bellwethers of change and how can they result in changing management procedures? Unforeseen, rapid changes can be observed everywhere, but are hardly a part of current management strategies.

Without including the human dimension in resource extraction and climate change, sustaining Arctic ecosystems and the people of the Arctic will remain an exorbitant goal. In order to strive for stability and predictability, it is of general importance that the 'science-society-management-alliance' includes the concept of resilience in all endeavours. ■

Resilience assessments and tipping points

A resilience assessment builds on the notion that humans and nature form an intertwined social-ecological system where the capacity to remain in a desired state for humans is determined by the resilience of the system. The combined interactions between different drivers can have greater impact on valued ecosystem services than each driver alone. Sometimes, change is gradual and the system will move forward in a predictable manner. However, as ecosystems and social systems are complex adaptive systems, change may also be non-linear. There are limits to how much shocks and disturbances a system can cope with and still recover to the desired state. Beyond those limits – i.e. tipping points – it functions differently. It either breaks down and encounters a period of recovery, or the system changes to some other functional state, after having reached a specific tipping point – a point of no return.

We are witnessing an emerging but still limited understanding of tipping points in the Arctic. In marine systems, observations point to dramatic changes in both ice cover and the underlying water column, with implications for species and productivity. This has already been addressed in some detail. For terrestrial ecosystems, evidence has accumulated of relatively rapid shifts in keystone plant communities throughout the Arctic. Vanishing permafrost alters the hydrology, accelerating the observed changes in ecosystem community composition and diversity. Rapid changes at the ecosystem and landscape levels in the tundra, such as increasing shrub coverage, are brought about by multiple drivers (such as temperature increase and permafrost release) and with feedbacks on the radiation balance adding to the surface warming. Ongoing landscape changes reduce or increase the carrying capacity for summer grazing by caribou and reindeer.

In addition, some studies have focused on social tipping points and community resilience in the Arctic, such as analyses of the economics of adaptation to climate change and the challenges for governance in turbulent times. Rather than assuming any deterministic relationships between changes in the physical and the social environment, an understanding of how ecosystem services influence the overall resilience of social-ecological systems, *including* attention to the ability of social systems to deal with various shocks and stresses, has to be a goal. ■

Multidisciplinary awareness: a precondition for resilience management

'The ink of the scholar is holier than the blood of the martyr.' 'The world subsists through the breath of school children.' 'The truth will set you free.' These are citations from three holy books, The Koran, the Talmud, and the New Testament, respectively. The three, closely interrelated, but alas so seriously fighting brethren of monotheism, thus agree about the fundamental importance of knowledge, wisdom of writings, and education. While this creates the base for all sustained progress and peaceful development, sustainability today demands more. In a world characterised by rapid change of its ecological and social systems it is as important to manage both culture systems and enhance their resilience as to manage the sustainable supply of specific products. Sustainability rests upon resilience management. In addition to basic knowledge, two additional dimensions are needed: multidisciplinary awareness and the human dimension.

This book on life in Svalbard, with its photographs, graphic evidence, and short texts on history, nature, climate, and humans, aims to contribute to multidisciplinary awareness, with an extraordinary archipelago at its focus. An immanent attentiveness to life in the Arctic that supports resilience management demands a wide range of knowledge. One has to bring together comprehension of climate and physical forcing, key species and ecosystems, resource exploitation and management, and vital political, ethnic, and cultural elements. Further, ethics and aesthetics have to be considered, as deficient aesthetics result in lack of respect for ethics. This is not a task for the talented hero, but, to the best of my knowledge, one for all of us. Resilience thinking depends upon a certain universality of knowledge, an ideal that materialised during the humanistic period and that our era has abandoned in its ever-accelerating search for more specialised information. The world transforms for the better not exclusively by the intellect, but only when man's wide-ranging qualities are involved and altered. We need a new, redefined humanism.

The suite of apparently superficially, little-connected aspects in the book have a paramount objective: to pave the ground for a more holistic, multidisciplinary comprehension that includes as much the human perspective as the imaginative mind. 'The rational mind is a faithful servant; the imaginative mind is a sacred gift', said Albert Einstein. A translation of the term Homo sapiens suggests that humans are wise. It may rather hint at that we have the capability to be wise. With the superb training of the rational mind that many of us are exposed to and that makes us one-sided, but not necessarily wise, we may ask how wisdom can be achieved. I suggest that multidisciplinary education, including imaginative exercising, art, and a wider variety of study directions, creates the precondition to tackle our future in a more adequate manner. It is this multidisciplinarity that creates the precondition for resilience management and sustainability maintenance. ∎

The beginning of the end?*

With multiple dangers lingering on the horizon, and inspired by Winston Churchill, one may be tempted to ask whether today's climate change is not the end? It is not even the beginning of the end! However, today's climate change is, perhaps, already the end of the beginning. This beginning will develop, if not adequately analysed through resilience thinking and counteracted by dedicated action, into a continuously reduced sustainability and carrying capacity. For many, it may make our planet a difficult place to live on in the future.

Future living conditions are and will be determined by the sustainable use of resources. It is not *the* or *a* future, but *our* future. We – all of us – *make* our future. The current destruction of our planet and the steady decrease of nature's carrying capacity (the base of our existence) is the dark, destructive, and generally unrecognised side of human civilisation. Human culture has continually decreased the carrying capacity of our planet since the very beginning. The following are a few examples: the Fertile Crescent, the cradle of human civilisation, became semi-desert; the rich European mixed woods became biodiversity deserts of steadily deprived farmland; the prairies and their abundant wildlife became monocultures; and the fertile Arctic and Antarctic seas have been deprived of large mammals. Hitherto, culture has been implicated in the decrease in nature's carrying capacity. The subjection and the enslavement of nature by human overpopulation, vividly exemplified in the Genesis of the *Old Testament* (which commanded: *Be fruitful and increase in number; fill the earth and subdue it*), is an exorbitant example of how man's destructive, nature-destroying mission is perpetually crammed into our minds. Was it really God's wish that Adam and Eve's descendants – who, for once, took him literally – should destroy his blue marble, Planet Earth? Today, we can no longer move into unexploited regions as we once did. The threat to the Arctic, with its rich resources, is that man's attitude to nature will probably result in large-scale habitat destruction, biodiversity losses, and pollution. Our proud human culture: why does this necessarily imply the ruin of our planet?

Many of us become saddened when confronted with these, alas so obvious, facts. However, there is no reason for melancholy and pessimism. Our planet has always been subjected to the cultural principles of *Homo sapiens*, i.e. to subjugate nature. Through this hubris, we enter increasingly a path of self-destruction. How can our culture remain what it endeavours to be: a positive, constructive, peaceful, and supportive contribution, a long-lasting spiritual home for humanity? As we today are increasingly aware of what we are doing, we now have the opportunity and moral duty to alter the current development. Do we want the negative trend and can we change it?

Carl Gustav Jung stated, that nature must not win the game, but she cannot lose it either. In the abyss of man's collective unconsciousness lingers both the snake Midgardsormen and the wolf Fenris of Norse mythology: ready to swallow the earth and the universe. The human race needs an 'ecosystem Freud'. In order to prevent us from self-destruction, humanity needs an indepth psycho- and eco-analysis. Such analyses are the base for a fruitful implementation of an understanding of the earth system. Have we ever thought about how we could achieve this? Where are the couches and who are the therapists? Moreover, who supervises the therapists? We know the symptoms, and we know the outcome, but where is the cure? Even more essential, where is the research to create the cure? ∎

---

* 'Science comes to a stop at the frontiers of logic, but nature does not – she thrives on ground as yet untrodden by theory'. (C.G. Jung ,1966)

Confidence in the future: a lost skill?

With the advent of our current vision of the universe through Copernicus, Kepler, and Newton, the balance, rootedness, and fundamental security of humans that resisted even throughout the upheavals of the medieval times, dissipated. Stars are no longer immutable, the earth is no longer the centre of the universe, and the universe is most likely of infinite extent and expanding. Heaven is far away, and hell is just beneath our feet. Kepler's *Harmonices Mundi* (The Harmony of the World) has faded away. How can we regain the all-embracing, fundamental, and self-centred confidence of the pre-science times, when all was moving, and relative, and physical reality was hardly imaginable?

The conflict of our recent times comprises a clash of opinions regarding the future and climate change. This cataclysm further weakens the confidence in the future. The majority of scientists throughout the world agree on the causes of the observed climate change and the effects it will have on human livelihoods. Scientists have the moral obligation to speak up and make knowledge public. This is one reason why society pays generously for advance in science. Why is it so difficult to accept the outcome of meticulous research? The perspective of politicians or that of company shareholders, i.e. a few years (e.g. related to re-election, margins, and yields), contributes to the rejection of Al Gore's campaign's call 'an inconvenient truth'. Both groups have to represent the people's or shareholder's wishes and needs. In the present era, such wishes are bound to the short-term perspective of a few years. This prevents politicians and economy managers from taking steps in favour of an ecologically sound and indispensible, longer time perspective. Does this suggest that our current time perspective is too restricted to ensure sustainability and resilience?

There are plenty of currently well-known popular politicians and economists that obtain high popularity by jumping onto the 'doubt-on-climate change' bandwagon. Finally, we have groups that always pop up in times of crises: those that propel doomsday scenarios onto the public stage. The hullabaloo of many consortia fills today's media space, and necessary decisions and adaptive steps are few. Meanwhile, climate change and negative ecosystem impacts carry on at full pace and in no place more intensely than the Arctic. Confidence in a predictable and secure future is what we all wish. Contemporary society seems to have lost this basic skill. We either become fearful about the future or we suppress the dark clouds on the horizon. According to Paul Valery, the problem of our time is that the future is not what it used to be.

How is knowledge of climate change communicated, understood, and accepted? Does it suggest that the public should look to the future with sufficient confidence to support a sustainable future? As man-made climate forcing proceeds and most probably will give rise to levels of global warming that are far higher than the official goals, most climate change scientists air what dangers are waiting for humanity during the remaining centennial. However, many people, if not the majority, think that the climate reflects natural variability and that all is how it always has been. How can this attitude be explained? It could probably be due to the interpretation of science messages: they can be incomprehensible or understood as fear messages. To keep an internal balance, fear messages are often rejected. Fear messages communicated by scientists or unbiased messages that are interpreted in a fright-evoking context do not support a balanced, realistic, and weighted assessment of the circumstances. To ensure that knowledge-based messages reach human hearts, science, in cooperation with the communication specialists, may have to develop an environmental communication psychology. Improved and realistic knowledge will probably have magic qualities: knowledge detoxifies over-dimensioned fear to cool facts.

The basic scientific concept of communication (ex cathedra), inspired by the arrogance that natural science is the only reliable pathway to make progress, is also recognised as the *knowledge deficit model*. The model has three phases: (1) scientists find the answers to a prominent question; (2) they communicate the knowledge to the people; and (3) the World instantaneously becomes a better place. Complaints by the people are misinterpreted as resistance to knowledge and change. If science wishes to have an effect on people's lives and to take action towards a more climate-friendly future, then science has to work hard on its self-perception. Science has to develop empathy for citizens and endurance to obtain change. Science has to foster a more human-friendly communication strategy. Determinism (a philosophy stating that for everything that happens there are conditions such that, given those conditions, nothing else could happen) and positivism (any system that confines itself to the data of experiments and excluding a priori or metaphysical speculations) can be strongly advocated by natural scientists. Some scientists do not seem to have taken note that both movements were defeated by the most advanced science already 100 years ago. Many a well-meaning scientist attempts to transform a world of choices and liberties into a deterministic jail. People react against this imprisonment of human self-determination.

Communication with citizens is today often called outreach. Without any doubt, to reach the joint goal of a sustainable world, science has to reach out. We have not only to meet the brains of the educated, but also the hearts of all contemporaries. In short, science has to become more humane. ■

The serenity of the mind

How can humans advance from the destructiveness of today's deadlock? When bearing in mind the relationship of humans and nature, it is important to consider the liaison as a system. A system is a collection of interdependent parts, enclosed within a defined boundary. The Earth System is the conglomerate formed by human civilisation and its planetary matrix, i.e. all parts of the earth that interact with humans and their activities. To consider the human and biophysical domains in isolation implies to come up with a partial solution that can lead to bigger problems, rather then solving them. The subjugation-of-nature strategy is a death trap, into which humanity has manoeuvred itself. Resilience – a system's capacity to adsorb disturbances without human hostile regime shifts – is the key to the sustainability that our planet desperately requires.

We need to create the mind space for resilience thinking. Man has to step down from attempting to be the master of nature, and science needs to descend from the pulpit of a self-declared 'truth doctrine'. It is an important step to meet the wider congregation of man in manners where all are equal. One first step could be that the Anglo-Saxon cultures modify the use of the term science, which is designated for a fraction of science: natural science. The application of the term reflects arrogance, dominance, and indoctrination that natural science tragically often propagates. 'Real science is natural science' is a common theme song. One gets the impression that natural science (unintended?) may have been inspired by Jesus: 'I am the way and the truth and the life' in John (14:6).

What will happen to our planet and mankind in decades to come? Resources are not sufficient for our present lifestyle and population increment. The carrying capacity of the world is decreasing due to overexploitation and the lack of strategies to base ways of living on renewable resources. There will be a good life of modest means also in the future, albeit different from the life we live today. However, to be optimistic instead of pessimistic demands that we shape our future in the best possible manner, with a serene but realistic attitude. That implies first of all that we have to change many of our traditional approaches and that we accomplish a thorough self-realisation analysis in an Earth System perspective. Humans are not the masters of nature as our traditional, but failing struggles suggest, but an integrated part of the earth's ecosystem. To achieve an Earth System perspective we seriously need more insight than natural science alone can provide.

Humans have to define themselves as partners of nature. Humanity has to be more humble concerning the modes that mankind has developed to explain reality. Natural science provides an unbeaten prolific mode without which we cannot exist anymore, but it is only one among quite a few approaches. Mankind needs time perspectives that are tuned to those of the ecosystems we depend upon. Only in this manner can we really act responsibly for a succession of future generations. We need to replace repression of undesirable realities or depressive mind-sets with realistic optimism and a serene mind.

The following contemplation by Reinhold Niebuhr may come close to our challenge: 'Grant to us the serenity of mind to accept that which cannot be changed; courage to change what can be changed, and wisdom to know the one from the other.' ■

XI: FAREWELL TO SVALBARD

We have arrived at the end of the book. Regardless how often we visit the archipelago or how long we live there, we are all visitors in Svalbard, which can never become our homeland. After exploring the landscapes and vistas, delving into and sharing experiences of life in Svalbard, all bid their farewell. Thus, sooner or later we head south from our sojourns in this Arctic archipelago – the last bastion of a fading human civilisation on the verge of the oblivious expanse of the Arctic Ocean. Those that are infected by the Arctic bug are, again and again, driven northwards by strong inner forces, like migratory birds that have to search for the northern skies in spring. Entering the realm of prehistoric forces and touched by the numinous, the power or presence of an archaic divinity, we become overwhelmed by speechless awe and immense longing. Despite of all its concreteness, rocks, ice, and life-threatening climate, we leave Svalbard as a dream-like stage set, onto which we project essentials of our soul. Svalbard becomes a space- and timeless backdrop of our existence.

With sadness we see the desolation disappear in the haze, where once again we have spent eleven months with gales, frost, and drifting snow. We encountered days and nights that were inordinately, excessively stunning, surrounded by silence that literally tormented our eardrums. The land that we abandon, with its summits and glaciers, is not as welcoming as Norway in summer, with woods that whisper at dusk, but nevertheless we exchange places with grief in our hearts. Those that have lived up there and understand the land, sense an inexplicable longing in their soul, and they are drawn back.
They have not only seen storms, when the sea with unrestrained force threw itself against the cliffs, burying all along its path. No, they have also witnessed the quiet evenings when land and sea are set on fire by blazing sunshine, when the mountain slopes reverberate with the sounds of chirping, and when tiny red and white flowers sprout and endeavour to cover the naked cliffs.
(W. Wolstad 1956, translated)

Arctic nature strongly impacts the human mind. It lures the thoughts back to the beautiful, lonely tranquillity and to a profound sense of nature. The language of life there, in its archetypical beauty, talks to one's soul. These experiences provide a feeling of harmony for those individuals that seek the High North in bright summer. When they return towards the southern regions, where life unfolds in a wide variety of forms, but where circumstances of dependency are restraining, their thoughts will once again longingly seek northwards: to the land of light, with its few, but harmonic lifestyles.
(H. Resvold-Holmsen 1930, translated)

Photo credits and captions

In addition to those mentioned in the Introduction and listed below the following persons have been essential for access to the graphics and photographs in the book: Susan Barr, Martin Berg, Katja Eklund, Gro Hørthe, Anders Frederik Jynge, Bill Krog, Anna Pasternak, Isabelle Messerli, Guro Rønningsgrend, Leiv Rosengren, Henrik Somdal, Petar Tale, Per Arne Tollefsen, Jarle Tveter, Anne-Catherine Biedermann and Marie Wefring.

from left to right and top to bottom

CHAPTER I. PRELUDE

Page 2-3
Kapp Lee, Svalbard, 2007
Photo: Rudi Caeyers

Page 8
Portrait of Albert I, Prince of Monaco
Photo: Nadar
Courtesy Wikipedia

Page 11
Isbre
Louis Tinayre, 1906
Courtesy Norwegian Polar Institute

Page 14-15
Negribreen
Photo: Arvid Sveen

Page 16
Fram Strait, Arctic Tipping Points (ATP)-cruise, 2011
Photo: Rudi Caeyers - BFE/UIT

Page 18
Skygge i blått rom
Johanne Marie Hansen-Krone

Page 21
Kåfjorden
Photo: Arvid Sveen

Page 22
KV Svalbard, iAOOS-cruise, 2008
(integrated Arctic Ocean Observing System)
Photo: Rudi Caeyers

Page 25
Midnight sun over Fram Strait, iAOOS-cruise, 2007
Photo: Rudi Caeyers - BFE/UIT

Page 26
Fridtjof Nansen, 1920

Page 28-29
Helicopter deck on-board KV Svalbard, iAOOS-cruise, 2007
Photo: Rudi Caeyers - BFE/UIT

CHAPTER II. THE SVALBARD ARCHIPELAGO AT A GLANCE

Page 30-31
Kullkaia, Longyearbyen, ATP-cruise, 2010
Photo: Rudi Caeyers - BFE/UIT

Page 32
Poli Arctici, map by Fredericum de Wit, Amsterdam
Courtesy Norwegian Polar Institute

Page 34
Map of Svalbard, Bjørnøya and Europe

Page 36
Atlantic and Arctic currents in the Arctic Ocean
Courtesy Norwegian Polar Institute

Page 36-37
Billefjorden, Svalbard, ATP-cruise, 2011
Photo: Rudi Caeyers - BFE/UIT

Page 38-39
Hornsund alps, ATP-cruise, 2011
Photo: Rudi Caeyers - BFE/UIT

Page 40-41
Hornsund foggy mountains, ATP-cruise, 2010
Photo: Rudi Caeyers - BFE/UIT

Page 42-43
Smeerenburg glacier, ATP-cruise, 2011
Photo: Rudi Caeyers - BFE/UIT

Page 44-45
Glacier ice, Magdelenefjorden, ATP-cruise, 2011
Photo: Rudi Caeyers - BFE/UIT

Page 47
Photo: Arvid Sveen

Page 48-49
NW Svalbard, ATP-cruise, 2011
Photo: Rudi Caeyers - BFE/UIT

Page 50-51
Isfjorden, iAOOS-cruise, 2008
Photo: Rudi Caeyers

Page 52-53
North Svalbard, ATP-cruise, 2011
Photo: Rudi Caeyers - BFE/UIT

Page 54
Marginal ice zone, Svalbard,
Photo: Rudi Caeyers - BFE/UIT

Page 55
Left: remains of a seal, Svalbard, ATP-cruise, 2009
Right: West Svalbard, ATP-cruise, 2010
Photo: Rudi Caeyers - BFE/UIT

Page 56
Polar bear and a flock of Little Auks, ATP-cruise, 2010
Photo: Rudi Caeyers - BFE/UIT

Page 57
Top: Harp seal
Bottom: flock of seals, ATP-cruise, 2009
Photo: Rudi Caeyers - BFE/UIT

CHAPTER III. SVALBARD HISTORY: THE EARLY DAYS

Page 58-59
Prins Karls Forland, Svalbard, 2005
Photo: Rudi Caeyers

Page 60
Spitsberga, map
Courtesy Norwegian Polar Institute

Page 62
The Death of Willem Barentsz (1836) by Christiaan Julius Lodewyck Portman
Courtesy Royal Museums, Greenwich

Page 65
Walvisvangst, Abraham Storck, Amsterdam 1654
Courtesy Rijksmuseum Amsterdam

Page 66
Première vue en panorama de Bell-Sound,
Auguste Mayer, La Recherche expédition, 1838-1840
Courtesy Tromsø University Museum

Page 69
Regiones SVB Polo Artico, map
Courtesy Norwegian Polar Institute

Page 70
Glacier a aiguilles et mouillage de la Recherche à Bell-Sound
Auguste Mayer, La Recherche expédition, 1838-1840
Courtesy Tromsø University Museum

Page 71
Plate LXVII from Sars, G. O., (1900). *An account of the Crustacea of Norway: Cumacea*, Bergen Museum.
Courtesy Wikimedia Commons

Page 73
6ᵉ vue en panorama de la baie de la Madeleine
Barthélemy Lauvergne, La Recherche expédition, 1838-1840
Courtesy Tromsø University Museum

Page 74
Portrait of Elling Carlsen
Photo: J.M. Jacobsen
Courtesy Norwegian Polar Institute

Page 76
Illustration of artifacts collected by later expeditions to the wintering site of Willem Barentsz, 1873. From *Relikwieën uit onzen Heldentijd: De Aarde en haar volken*
Courtesy Wikipedia

Page 77
Nova Zembla flute
Courtesy Rijksmuseum Amsterdam

CHAPTER IV. FROM THE EDGE OF CIVILISATION TO A MODERN HUB

Page 78-79
Kullkran Panorama, Longyearbyen, 2011
Photo: Rudi Caeyers - BFE/UIT

Page 80
Photo: Francois Lenoir / Reuters / NTB scanpix

Page 82-83
Pictures taken at the Polar Museum, Tromsø, 2012
Photo: Rudi Caeyers - BFE/UIT

Page 84
Atlantis at Spitsbergen
by Kenneth D. Shoesmith,

Page 87
No title
by Gert Jynge
Courtesy Norsk Bergverksmuseum, Kongsberg

Page 88
Spitzberg, Baleinerie de Green Harbour
by Louis Tinayre, 1907

Page 91
Excerpt from Svalbard Treaty
Courtesy Norwegian Polar Institute

Page 92
WWII: Norwegian soldier in Longyearbyen after German attack, Spitsbergen 8 September 1943.
Photo: NTB Scanpix

Page 93
WWII - Norwegian soldiers in Longyearbyen.
Photo: NTB Scanpix

Page 95
Nybyen and Sverdrupbyen in Longyearbyen, Svalbard 1950
Photo: Leif Pedersen
Courtesy Norwegian Polar Institute

Page 96-97
Longyearbyen Airport, 2012
Photo: Rudi Caeyers - BFE/UIT

Page 98-99
Tourists and cruise ship
Photo: Camille Seaman

Page 100-101
UIT's research vessel, formerly called Jan Mayen, now Helmer Hanssen
Framstrait, ATP-cruise, 2011
Photo: Rudi Caeyers - BFE/UIT

Page 102
University in Svalbard (UNIS), interior.
Photo: Camille Seaman

Page 102-103
Ny-Ålesund, 2005
Photo: Rudi Caeyers

Page 104
Children's party, Longyearbyen
Photo: Svein Holo / NN / Samfoto / NTB scanpix

Page 105
From left to right and from top to bottom:
Wine cellar, photo: Espen Sjølingstad Hoen / VG / NTB scanpix
Sun Day, photo: Berit Roald / NTB scanpix
Svalbard limousine, photo: Arvid Sveen
Miner, photo: Berit Roald / NTB scanpix
Kebab car Røde isbjørn, photo: Camille Seaman
Shoes, photo: Berit Roald / NTB scanpix
Art, photo: Arvid Sveen
Biker, photo: Berit Roald / NTB scanpix

CHAPTER V: NATURE
Page 106-107
Collection of micro algae
Courtesy University of Tromsø, BFE-faculty and professor Hans Christian Eilertsen

Page 108
Watercolors of zooplankton by Miguel Medrano Alcaraz

Page 111
Kunstformen der Natur (1904), plate 1: Phaeodaria
By Ernst Haeckel
Courtesy Wikipedia

Page 112
Marine bacteria
Photo: professor Gunnar Bratbak
Courtesy University of Bergen

Page 114
Skeletonema marinoi
Photo: Maria Degerlund
Courtesy University of Tromsø, BFE-faculty

Page 114-115
Underwater view of pack ice with clouds of ice algae.
STENZEL, MARIA/ National Geographic Stock

Page 116
Copepod
Photo: Joan Costa

Page 118-119
Krill
Photo: Joan Costa

Page 120-121
Amphipod
Photo: Joan Costa

Page 123
Kunstformen der Natur (1904), plate/planche 49: Actiniae
By Ernst Haeckel
Courtesy Wikipedia

Page 124-125
Photos: Peter Bondo Christensen

Page 126-127
Photo: Erling Svensen / UWPhoto

Page 128
Photo: Erling Svensen / UWPhoto

Page 130-131
Shrimp
Photo: Rudi Caeyers - BFE/UIT

Page 132-133
Catch, ATP-cruise, 2010
Photo: Rudi Caeyers - BFE/UIT

Page 134
Polar cod
Photo: Peter Leopold

Page 135
Capelin
Photo: Arve Lynghammar

Page 136
Cod, ATP-cruise, 2010
Photo: Rudi Caeyers - BFE/UIT

Page 136-137
Cod
Photo: Erling Svensen / UWPhoto

Page 138-139
Keenpress/National Geographic Stock

Page 140
Polar bear, feeding on a seal carcass
Photos: Camille Seaman

Page 141
Polar bear, ATP-cruise, 2009
Photo: Rudi Caeyers - BFE/UIT

Page 142
Ringed seal
Photos: Bjørn Frantzen / Norsk Polarinstitutt

Page 143
Bearded seal
Photo: Audun Rikardsen

Page 144
Walrus
Photo: Audun Rikardsen

Page 145
Left photo: Camille Seaman
Right photo: Rudi Caeyers - BFE/UIT

Page 146
Minke whale in midnight sun
Photo: Audun Rikardsen

Page 147
Fin whale
Photos: Fredrik Broms

Page 148-149
Bjørnøya, Miseryfjellet
Photo: Arvid Sveen

Page 150-151
Arctic fox
Photos: Fredrik Broms

Page 152
Reindeer
Photo: Espen Bergersen

Page 153
Photo: Camille Seaman

Page 154-155
Bird colony on Hopen, Svalbard, ATP-cruise, 2010
Photo: Rudi Caeyers - BFE/UIT

Page 156-157
Little auk and Arctic tern
Photos: Espen Bergersen

Page 158-159
Glaucous gull and Northern fulmar
Photos: Espen Bergersen

Page 160-161
Snow bunting and Kittiwake
Photos: Espen Bergersen

CHAPTER VI: COMMUNITIES, THEN AND NOW

Page 162-163
Pyramiden, Svalbard, ATP-cruise, 2011
Photo: Rudi Caeyers - BFE/UIT

Page 164
Lenin statue at Pyramiden, Svalbard, Outreach-cruise, 2012
Photo: Rudi Caeyers - BFE/UIT

Page 166
Advent Bay
Painting of Longyearbyen by Loius Tinayre, 1907
Courtesy Musée océanographique de Monaco

Page 166-167
Longyearbyen
Photo: Arvid Sveen

Page 168
University Centre in Svalbard, Outreach -cruise, 2012
Photo: Rudi Caeyers - BFE/UIT

Page 169
Top: early settlements of Longyearbyen
Courtesy Musée océanographique de Monaco
Bottom: remains of foundations for housing, Svalbard, 2005
Photo: Rudi Caeyers

Page 170-171
Pyramiden, Outreach-cruise, 2012
Photos: Rudi Caeyers - BFE/UIT

Page 172-173
Pyramiden, Outreach -cruise, 2012
Photos: Rudi Caeyers - BFE/UIT

Page 174-175
Barentsburg, 2005
Photo: Rudi Caeyers

Page 176-177
Barentsburg, Outreach -cruise, 2012
Photos: Rudi Caeyers - BFE/UIT

Page 178
Smeerenburg, Outreach -cruise, 2012
Photo: Rudi Caeyers - BFE/UIT

Page 179
Left: *Vue prise dans la baie de Smeremberg*
Barthélemy Lauvergne, La Recherche expédition, 1838-1840
Courtesy Tromsø University Museum
Right: Smeerenburg, Outreach -cruise, 2012
Photo: Rudi Caeyers - BFE/UIT

Page 180-181
Smeerenburg, Outreach -cruise, 2012
Photos: Rudi Caeyers - BFE/UIT

Page 182-183
Virgohamna, Outreach -cruise, 2012
Photo: Rudi Caeyers - BFE/UIT

Page 184
Left: Wellmann's station, Virgohamna, 1909
Right: Wellmann's first attempt to reach the North Pole with airship America
Courtesy Norwegian Polar Institute

Page 185
Virgohamna today, Outreach -cruise, 2012
Photo: Rudi Caeyers - BFE/UIT

Page 186-187
Ny-Ålesund airport, Svalbard, Outreach-cruise, 2012
Photo: Rudi Caeyers - BFE/UIT

Page 188
Top to bottom:
Ny-Ålesund, cable car to Zeppelin station, 2005
Station roof, with sniffers and Kim Holmén, Zeppelin station manager, 2005
Photos: Rudi Caeyers

Svalbard summer, Kings Bay
Painting by Gunnar Wefring
Courtesy Nordøsterdalmuseet, Hedemark fylkeskommune AS

Top right: Airship Norge, 1926
Courtesy Norwegian Polar Institute

Page 189
Glacier ice, Kings Bay, Svalbard, 2005
Photo: Rudi Caeyers

Page 190-191
Left: New-London, remains of the Northern Exploration Company Ltd., Svalbard, 2005
Top right: desintegrated marble after defrosting, Kings Bay, Svalbard, 2005
Bottom right: New-London, Camp Mansfield, Svalbard, 2005

Page 192-193
Svea mine and town, Svalbard, 2007
Photos: Tommy Dahl Markussen

Page 194-195
Hornsund panorama
Bottom: Polish station at Hornsund, Svalbard, ATP-cruise, 2009
Photos: Rudi Caeyers - BFE/UIT

Page 196-197
Hornsund glacier, Svalbard, ATP-cruise, 2010
Photo: Rudi Caeyers - BFE/UIT

Page 198
Hopen beach and bird colony, Svalbard, ATP-cruise, 2010
Photo: Rudi Caeyers - BFE/UIT

Page 199
Hopen station interior, Svalbard, ATP-cruise, 2010
Photo: Rudi Caeyers - BFE/UIT

Page 200
Hopen, one-blue-eyed resident, Svalbard, ATP-cruise, 2010
Photo: Rudi Caeyers - BFE/UIT

Page 201
Hopen station manager, Kåre Holter Solhjell, Svalbard, ATP-cruise, 2010
Photo: Rudi Caeyers - BFE/UIT

Page 202-203
Bjørnøya, Svalbard, ATP-cruise, 2009
Photos: Rudi Caeyers - BFE/UIT

Top right: bird colony, Bjørnøya, Svalbard
Photo: Camille Seaman

CHAPTER VII: PEOPLE, THEN AND NOW

Page 204-205
Weather balloon launch, Ny-Ålesund, Svalbard, Outreach-cruise, June 2012
Photo: Rudi Caeyers - BFE/UIT

Page 206
Pyramiden bar, Svalbard, Outreach-cruise, 2012
Photo: Rudi Caeyers - BFE/UIT

Page 208
Cruise participants: Louis Tinayre (second from left), Gunnar Isachsen (centre), and Prince Albert I (to the right)
Courtesy Musée océanographique de Monaco

Page 209
Prince Albert I of Monaco
Courtesy Musée océanographique de Monaco

Page 210
Balloon house at Virgohamna,
Photo: Ludvig Blunck
Courtesy Norwegian Polar Institute

Page 211
Portrait of Salomon August Andrée, photo from the atelier of Gösta Florman in Stockholm.
Courtesy Grenna Museum - Polarcenter

Page 213
Portrait of Amundsen, 1935, by Astri Welhaven Heiberg
Photo: Rudi Caeyers
Courtesy Norwegian Polar Institut

Page 214
Vue de l'Ile de l'Ours ou Beeren-Eiland
François-Auguste Biard, La Recherche expédition, 1838-1840
Courtesy Tromsø University Museum

Page 217
Portrait of Fridtjof Nansen, 1938, by Erik Werenskiold
Photo: Jan Dalsgaard Sørensen/FNI
Courtesy Polhøgda, Oslo

Page 218
Portrait of Gunnerius Isachsen, 1906, by Louis Tinayre
Photo: Rudi Caeyers
Courtesy Norwegian Polar Institute

Page 219
Left: support ship *Kvedfjord* from Elde, Kvæfjord.
Right: Gunnar Isachsen onboard the research ship *Princesse Alice II*
Courtesy Musée océanographique de Monaco

Page 220
Portrait of Adolf Hoel, 1936, by Gunnar Wefring
Courtesy Svalbard Museum

344 | Svalbard Life

Page 221
La mission Isachsen engagé sur un glacier franchit un port de glace, 1906
By Louis Tinayre
Courtesy Musée océanographique de Monaco

Page 222-223
Scientists returning from ice core drilling, iAOOS-cruise, 2008
Photo: Rudi Caeyers

Page 224
Portrait of Ivan Starostin

Page 225
Portrait of Hilmar Nøis
Photo: Th. Winsnes
Courtesy Norwegian Polar Institute

Page 227
Portrait of Léonie d'Aunet, 1842
By François-Auguste Biard
Photo: RMN – Daniel Arnaudet
Courtesy Châteaux de Versailles et de Trianon

Page 228-229
Panorama of a glacier at Magdelenefjorden, Svalbard, ATP-cruise, 2009
Photo: Rudi Caeyers - BFE/UIT

Page 230
Portrait of Hanna Resvoll-Holmsen at Blomstrands havn, 1908
Photo: Adolf Hoel
Courtesy Norwegian Polar Institute

Page 233
Portrait of Wanny Woldstad
Courtesy Norwegian Polar Institute

Page 234
Coal
Photo: Rudi Caeyers

Page 235
Portrait of 3 miners, 1986
By A. Strakhov
Courtesy Barentsburg Museum

Page 236-237
Pyramiden panorama, Svalbard, Outreach-cruise, June 2012
Photo: Rudi Caeyers - BFE/UIT

Page 238-239
Sysselmannsbygget, Longyearbyen, Svalbard, 1998
Photo: Arvid Sveen

Page 239
Portrait of Odd Olsen Ingerø, governor of Svalbard, ATP-cruise, June 2011
Photo: Rudi Caeyers - BFE/UIT

Page 240-241
From left to right:
Portrait of Lena Romanenko, Barentsburg, 2011
Portrait of Roger Jacobsen, Ny-Ålesund, 2011
Portrait of Hans Roar Hansen, Svalbard, 2011
Portrait of Fiona Danks, Ny-Ålesund, 2011
Portrait of Carlos Duarte, Svalbard, 2011
Portrait of Tove Gabrielsen, Longyearbyen, 2012
Portrait of Ole Magnus Rapp, Svalbard, 2011
Portrait of Jan Martin Berg, Svalbard, 2011
Portrait of Johnna Holding, Svalbard, 2011
All photos: Rudi Caeyers - BFE/UIT

CHAPTER VIII: SVALBARD LIFE IN ART

Page 244-245
Svalbard Triptych I, 2012
Charcoal and pastel on Ingres, by Rudi Caeyers

Page 246
Kullkaia, 2012
Charcoal and pastel on Ingres, by Rudi Caeyers

Page 249
Onboard "Laila", Wigdehl paints Alkhornet, 1910
Photo: Staxrud
Courtesy Norwegian Polar Institute

Page 250
Magdalena Bay, 1839
Oil painting by François-Auguste Biard
Courtesy Musée du Louvre, Paris

Page 251
Fight with polar bears
Oil painting by François-Auguste Biard
Courtesy Nordnorsk Kunstmuseum, Tromsø, Norway

Page 252-253
Vue prise dans la baie de la Madeleine
Lithography by Barthélemy Lauvergne, La Recherche expédition, 1838-1840
Courtesy Tromsø University Museum

Page 254
Morgen ved Norskeøyne, 1879
Oil painting by Franz Wilhelm Schiertz
Courtesy Nasjonalmuseet, Oslo, Norway

Page 255
Untitled?, 1880
Watercolour by Franz Wilhelm Schiertz
Courtesy Norwegian Polar Institute

Page 256
Erling Yarl im Packeis, 1896
Oil painting by Hans Beat Wieland
Courtesy Stiftung für Kunst, Kultur und Geschichte, Winterthur, Switzerland. Inv. Nr. 4370

Page 257
Danske Bay, Spitzbergen, 24./25. Juli 1896
Oil painting by Hans Beat Wieland
Courtesy Bill Krog, Forch, Switzerland

Page 258
Glacier Post, Sassen Bay, Spitzberg 1898
Watercolour by Witold Lovatelli Colombo
Courtesy Musée Océanographique de Monaco

Page 259
Mont Temple, Baie Sassen, Spitzberg 1898
Watercolour and mixed media by Witold Lovatelli Colombo
Courtesy Musée Océanographique de Monaco

Page 260
Untitled, 1907
Pastel and mixed media, by Louis Tinayre
Courtesy Norwegian Polar Institute

Page 261
La Baie et le Glacier Louis Tinayre, 1906
Oil painting by Louis Tinayre
Courtesy Musée Océanographique de Monaco

Page 262
Left: *Front du glacier de la Baie Ginevra, 1898*
Right: *Rue de glace au Glacier Post, 1898*
Oil paintings by Marius Borrel
Courtesy Musée Océanographique de Monaco

Page 264
Vestisen, 1882
Watercolour on photo, by Fridtjof Nansen
Courtesy Norwegian Polar Institute

Page 265
Kvinne
Lithography, by Fridtjof Nansen
Courtesy Norwegian Polar Institute

Page 266-267
Left: *Untitled - right: Untitled, 1919*
Oil paintings by Michaloff Wigdehl
Courtesy Norwegian Polar Institute

Page 268
Untitled
Watercolour
Courtesy Norsk Bergverksmuseum

Page 269
Kullgruve, 1932
Oil painting by Gerd Jynge
Courtesy Nordnorsk Kunstmuseum, Tromsø

Page 270
Untitled, 1942
Oil painting by Heinz Köhler
Courtesy Svalbard Museum

Page 271
20h am 17. Mai, Mitra, 1942
Oil painting by Heinz Köhler
Courtesy Svalbard Museum

Page 272
Barentsburg, 1986
Oil painting by Andrei Alekseevich Yakovlev
Courtesy Barentsburg Museum

Page 273
Mine Pyramiden, 1986
Oil painting by Andrei Alekseevich Yakovlev
Courtesy Barentsburg Museum

Page 274
Arktisk månenatt, 19xx?
By Kåre Tveter
Courtesy National Gallery, Oslo

Page 276-277
Left: *Isfjorden, Svalbard*
Oil painting by Kåre Tveter
Right: *Isfjorden, Svalbard*
Aquarelle by Kåre Tveter
Courtesy Petar Tale and Gro Hørthe

Page 278
Left: *"DIVISO III-XI" 2011*
Right: *"DIVISO V-XI" 2011*
Sculptures by Terje Roalkvam

Page 279
"DIVISO III-XI" 2011
Sculptures by Terje Roalkvam

Page 280
Kongsbreen
Oil painting by Vemund Thoe

Page 281
Fram
Oil painting by Vemund Thoe

CHAPTER IX: CLIMATE CHANGE IN THE ARCTIC: WHAT IS AHEAD OF US?

Page 282-283
Fram Strait panorama 3, 2008
Photo: Rudi Caeyers

Page 284
One small step for man, iAOOS-cruise, 2007
Photo: Rudi Caeyers - BFE/UIT

Page 286
Anchoring KV Svalbard to the ice sheet, iAOOS-cruise, 2008
Photo: Rudi Caeyers

Page 288-289
Watercloud over the Fram Strait, iAOOS-cruise, 2008
Photo: Rudi Caeyers

Page 290-291
Fuel depot, Agardh Bay, Svalbard, 1988
Photo: Arvid Sveen

Page 292-293
Sea ice, from helicopter, iAOOS-cruise, 2007
Photo: Rudi Caeyers - BFE/UIT

Page 294
Sea ice extent in the Arctic in September 1980 (top) and September 2012 (bottom)
Courtesy National Snow & Ice Data Center (nsidc.org)

Page 297
Sea butterfly (*Limacia helicina*)
Photo: Erling Svensen / UWPhoto

Page 298-299
Photo: Camille Seaman

Page 300
Tourists in the early days, 1897
Photo: Hans Beat Wieland

Page 300-301
Tourists on a Hurtigruten cruise, 2008
Photo: Camille Seaman

Page 302
Longyearbyen power plant, ATP-cruise, 2011
Photo: Rudi Caeyers - BFE/UIT

Page 305
Fishing cod
Photo: Arvid Sveen

CHAPTER X: OUTLOOK AND PERSPECTIVE
Page 306-307
Barents Sea, ATP-cruise, 2011
Photo: Rudi Caeyers - BFE/UIT

Page 308
Fogbow, ATP-cruise, 2009
Photo: Rudi Caeyers - BFE/UIT

Page 310-311
Barentsburg power plant, ATP-cruise, 2009
Photo: Rudi Caeyers - BFE/UIT

Page 312
V
iAOOS-cruise, 2007
Photo: Rudi Caeyers - BFE/UIT

Page 314-315
Wake of icebreaker KV Svalbard, iAOOS-cruise, 2007
Photo: Rudi Caeyers - BFE/UIT

Page 316-317
Scientist collecting sediment samples, iAOOS-cruise, 2007
Photo: Rudi Caeyers - BFE/UIT

Page 318-319
Kajak, Kullorsuaq, Greenland, Hurtigruten cruise, 2009
Photo: Rudi Caeyers

Page 320
Sisimiut youngster, Greenland, Hurtigruten cruise, 2009
Photo: Rudi Caeyers

Page 323
Garbage dump at Nuussuaq, Greenland, 2009
Photo: Rudi Caeyers

Page 324
Old machinery at Ny-London, Svalbard, 2005
Photo: Rudi Caeyers

Page 327
Bow spray, iAOOS-cruise, 2008
Photo: Rudi Caeyers - BFE/UIT

Page 328
Memorial plate of deceased Polish scientist and instrument tripod at Wilczekodden, Hornsund, Svalbard, 2010
Photo: Rudi Caeyers - BFE/UIT

Page 331
Foggy day in Fram Strait, iAOOS-cruise, 2007
Photo: Rudi Caeyers - BFE/UIT

Page 332
Knife, iAOOS-cruise, 2007
Photo: Rudi Caeyers - BFE/UIT

Page 335
Ice plains, iAOOS-cruise, 2007
Photo: Rudi Caeyers - BFE/UIT

Page 336-337
Drift ice off the Svalbard coast, ATP-cruise, 2011
Photo: Rudi Caeyers - BFE/UIT

Page 338
Bird eye perspective of Svalbard mountains and fjords in spring
Photo: Arvid Sveen ■

List over applied publications

Atlas of the Marine Fauna of Southern Spitsbergen (1990-1997). Vol. 1-2(a, b, c). Ossolineum, Wrocław, Poland. http://www.iopan.gda.pl/ekologia/strona6.html

Banerjee, S. (2012). Arctic voices. Resistance at the tipping point. Seven Stories Press, New York.

Baudelaire, C (1863). The Painter of Modern Life. In "The Painter of Modern Life and Other Essays", (translation Jonathan Mayne, 1964).

Bjerck, H.B., Johannessen, L.F. (1999). Virgohamn. I lufta mot Nordpolen. Sysselmannen på Svalbard. 1-35.

Bjørnæs, C., Prestrud, P. (2012). The state of the poles. Climate lessons from the International Polar Year. Unipub Norway, 1-139

Carpenter, E. (1973). Eskimo realities. Holt, Rinehart and Winston, New York.

Climate change in the Norwegian Arctic. Consequences for life in the north. 136 Report Series Norwegian Polar Institute, Tromsø, Norway, 1-135.

D´Aunet, L. (1854). Voyage d'une femme au Spitzberg. Hachette, Paris.

Davenport Adams, W. H. (1903). Celebrated Women Travellers of the ninetheenth century. E. P. Dutton & Co, New York.

Duarte, C, Wassmann, P (eds) (2011). Arctic Tipping Points. Fundacion FBBVA. http://www.fbbva.es/TLFU/microsites/artic/ATPweb.html

Dufferin, The Marquess of (1903). Letters from High Latitudes. The Merlin Press, London.

Evans, G. (2011). The Maritime Art of Kenneth D. Shoesmith. Silver Link Publishing Ltd, Peterborough, Great Britain

Fløgstad, K. (2007). Pyramiden. Portrett av en forlaten utopi. Spartakus Forlag, Oslo, 1-181.

Gogol, N. (1842). The Government Inspector. St. Petersburg.

Haeckel, E. (1904). Kunstformen der Natur. Bibliographisches Institut, Leipzig und Wien.

Jenssen, K (1979). Michaloff Wigdehl. Nordlands og Spitsbergens maler. Tiden Norsk Forlag, Oslo, 1-55.

Jung, C.G. (1966). The practice of psychotherapy: Essays on the psychology of the transference and other subjects. Collected Work Vol. 16. Princeton University Press. First published in 1954.

Kortsch, S., Primicerio, R., Beuchel, F., Renaud, P.E., Rodrigues, J., Lønne, O.J., Gulliksen, B. (2012). Climate-driven regime shifts in Arctic marine benthos. www.pnas.org/cgi/doi/10.1073/pnas.1207509109

Løvland, B., Sandbo, G. (2011). Hilsen fra Svalbard. Postkort – Frimerker – Poststempler. Dreyers Forlag, Oslo, 1-268.

Lopez, B. (1986). Arctic dreams. The Harville Press, 1-464.

Meløy, S. Sandberg (2012). Polarheltinner. Gyldendal, Oslo, 1-230.

Miller, A. I. (2001). Space, time, and the beauty that causes havoc. Basic Books, New York, 1-357.

Nansen, F. (1911). In Northern Mists: Arctic Exploration in Early Times. Vol. II. Fridtjof Nansen. London 1911.

Nansen, F. (1920). En ferd til Spitsbergen. Jacob Dybwads Forlag, Christiania, 1-280.

Pauli, W. (1955). The Influence of Archetypal Ideas on the Scientific Theories of Kepler. Routledge & Kegan Paul, London. Originally in: C. G. Jung & W. Pauli (1952). Naturerklärung und Psyche, Rascher Verlag, Zürich.

Poulsen, H.K., and Holm, H. (2009). Nature strikes back. Man and nature in western art. Statens Museum for Kunst, Copenhagen, 1-163

Lamont, J. (1861). Season with the sea-horses or Sporting adventures in the northern seas. Hurst and Blackett, London 1-312

Poincaré, H. (1902). Science and Hypothesis. NY, Dover 1952. Originally published by Flammarion in 1902.

Reilly, J. T. (2009). Greetings from Spitsbergen. Tourist at the Eternal Ice 1827 -1914. Tapir Academic Press, 1-227

Rockström, J., Klum, M. (2012). The human quest. Prospering within the planetary boundaries. Bokförlaget Langenskiöld, Stockholm, 1-319.

Ryall, A., Schimanski, J., Howlid Wærp, H. (2010). Arctic Discourses. Cambridge Scholar Publishing, 1-339.

Schibsted, A. (1909). Turistforeningens årbok 1909.

Schimanski, J., Theodorsen, C., Howlid Wærp, H. (2011). Reiser og ekspedisjoner i det litterære Arktis. Tapir Akademisk Forlag, 1-386

Stange, R. (2008). Spitsbergen – Svalbard. A complete guide around the arctic archipelago. 1-540

von Franz, M.-L. (2000). The Problem of the Puer Aeternus. 3rd Edition, Inner City Books, Toronto.

Voyage Pittoresque (2005). Reiseskildringer fra nord. Nordnorsk Kunstmuseum, 72 pp

Walker, B., Salt, D. (2006). Resilience Thinking. Sustaining ecosystems and people in a changing world. Island Press, 1-174.

Wassmann, P. (2011). Arctic Marine Ecosystems in an Era of Rapid Climate Change. Progress in Oceanography 90: 1-17.

Wassmann, P., Lenton, T. Arctic tipping points in the Earth System perspective. AMBIO 41(1):1-9.

Wassmann, P. (2008). Impacts of global warming on Arctic marine ecosystems and processes. In: Impactos del calentamiento global sobre los ecosistemas polares. Carlos M. Duarte (ed). Fundación BBVA, Spain, 111-138.

Woldstad, W. (1956). Første kvinner som fangsmann på Svalbard. Tanum, Oslo. (Svalbardminner nr. 31, Vågemot Miniforlag 2005).

Web sites:
Wikipedia, Sysselmann på Svalbard, Svalbard Museum, Norwegian Polar Institute ∎